Surviving the 21st Century

Julian Cribb

Surviving the 21st Century

Humanity's Ten Great Challenges and How We Can Overcome Them

 Springer

Julian Cribb
Canberra, ACT, Australia

ISBN 978-3-319-41269-6   ISBN 978-3-319-41270-2   (eBook)
DOI 10.1007/978-3-319-41270-2

Library of Congress Control Number: 2016947168

This Springer imprint is published by Springer Nature
The registered company is Springer International Publishing AG Switzerland

Dustjacket Reviews

"Julian Cribb brilliantly introduced the general public to the gigantic threat of global toxification in *Poisoned Planet*. Now he's done it again, taking on the entire existential threat to civilization. Absolutely everyone with an interest in humanity should read this clear, authoritative, scary book.

> – Paul R. Ehrlich, co-author of '*The Annihilation of Nature*'. Bing Professor of Population Studies Emeritus; President, Center for Conservation Biology; Department of Biology, Stanford University

"With astonishing breadth of knowledge and acute observational skills, Julian Cribb has given us a book that is a kind of report on the state of life on the planet. At the centre of life on earth, he tell us, is the creature known as homo sapiens – self-deceiver, degrader, destroyer, anything it seems but sapiens. And yet, if we peer through the gloom is that a spark we can just make out, the spark of wisdom?"

> – Professor Clive Hamilton, author *Requiem for a Species* and *Earthmasters*

"We've come a long way from our hunter/gatherer past, but how assured is our future?In this book, Julian Cribb argues that the continuation of the human story depends on what we do now and in the immediate future."

> – Nobel Laureate Professor Peter Doherty, University of Melbourne

"Cribb has delivered another clear-eyed and expansive look over the problems we face, inspiring in both its scope and scholarship, and again has tempered the sense of doom with well-defined, positive actions for us all, both as a society and as individuals. It is a systemic problem, and he provides the necessary systemic solutions – may they be widely read and acted upon!"

– Dr Mark Stafford Smith, Chair, Science Committee, Future Earth

"An overpopulated, resource depleted and environmentally wounded planet needs our urgent help. Julian Cribb provides timely and thoughtful answers."

– Major General the Hon. Michael Jeffery AO AC, former Governor-General of Australia

"Only rarely does someone write a "must read" book. This is one of them. Nothing is more important than to truly tackle to massive challenges facing the Earth and humanity itself".

– Professor David Lindenmayer AO, Australian Research Council Laureate Fellow

"This could be one of the most important books of the 21stC, particularly if enough people read it, understand the message embedded in the content and then act accordingly. This well-written and researched book is more than a catalogue of despair but rather it points out some very obvious actions that could take humanity on a more sustainable journey."

– Professor Graham Durant, Director, Australian National Science and Technology Centre

"This is the guide for our times, the overlapping hazards we prefer not to think about but must. Here is a magisterial summary that spares no comfort zones but does show what we need to do and, at last, how to do it".

– Dr Robyn Williams, Science Broadcaster, Australian Broadcasting Corporation

"The structure of this book provides a fascinating device for exploring the great crises of our time, and for facing up to the biggest question: are we capable of dealing with them?"

– Bill McKibben, author of *The End of Nature* (1989) and *Eaarth* (2010)

Today we appear to be facing an increasingly uncertain future and are probably more confused than ever. Julian Cribb's book adds to these feelings but also provides glimmers of hope as he articulates with clarity what our challenges are and how we might confront them.

- Maj Gen John Hartley AO, former Australian Army Land Commander and Director of Defence Intelligence.

"This book concisely summarises the critical challenges facing human society in the twenty-first century, as well as providing helpful advice about the most useful steps individuals can take. It is comprehensive, accurate and measured in its assessments. It is an essential guidebook to help thoughtful people act responsibly."

- Emeritus Professor Ian Lowe, Environmental scientist, Griffith University

"The material in the book is exceptionally thoroughly researched and referenced; the author is a very distinguished science writer. The book is encyclopedic in scale. Everyone who wishes to be well-informed on the ills of civilization and how they might be solved, should read this book - particularly those in public office."

- Emeritus Professor Adrian Gibbs, Virologist, Australian National University.

"This erudite and highly readable analysis of the interlinked threats to the future of the human species is absolutely essential reading for all politicians and policy makers, voters and young people everywhere. Cribb shows with absolute clarity that humanity in the 21st-century now faces the greatest test of our collective wisdom in our relatively short history.Grandparents should read the book with particular care."

- Emeritus Professor Bob Douglas, Epidemiologist, Australian National University

This is an important book. Few others deal with so many confronting problems in an integrated way. Hopefully it will fulfil its aim of helping build the discussion about survival that we have to have.

- Jenny Goldie, past president, Sustainable Population Australia

In his latest book, "Surviving the 21st Century", Julian Cribb provides a masterful evidence-based account of the ten greatest threats to humanity – and importantly, how to beat them. This ground-breaking and timely treatise goes far beyond simply documenting gloom-and-doom to show how we can collectively achieve solutions to the world's major challenges.

- Distinguished Professor Terry Hughes, coral reef scientist, James Cook University

*This book is dedicated
to the memory of the late Professor Tony McMichael
(1942–2014), author of 'Planetary Overload',
and to Professor Paul R. Ehrlich,
author of 'The Population Bomb'.
They were right.*

*It is also dedicated to my granddaughter Vivienne,
whose generation must face the daunting challenge
of restoring our world.*

Preface

This book is about the future survival prospects of our species, *Homo sapiens,* in the twenty-first century.

It deals with the compound challenge of the ten greatest threats to our existence we humans have faced in the past million years—and what we can sensibly do about them.

It presents fresh evidence, from trusted scientific sources, to shine a light on the nature of the risks our vast numbers and overwhelming demands on the Planet are bringing upon us. It explores our strengths and weaknesses as a species in facing them.

These challenges are now so profound that I am constantly meeting people, all over the world—scientists, grandparents, young people—who are pessimistic, if not despairing, about the future we are leaving to our children, and to their children.

But this isn't simply a book about problems. It's absolutely about solutions. It is about hope—though a hope that is well founded, on fact and science, not simply on belief, ignorance or wishful thinking.

It's about how we humans can arrive at a common, clear insight into the nature of our greatest test—and into our own natures—in order to work together as a species to solve it and prosper.

It's about the very thing we humans have always done best: understand and find co-operative solutions to life-threatening challenges.

In exploring our greatest risks, this book does not pretend to predict the future. That's not possible. It simply presents the best evidence, arguing that a sound awareness of those risks gives us greater influence over our future and the options we have for shaping it. Each chapter concludes with clear advice on what can be done at global level but also what individuals can do for themselves to make a global difference. Importantly, it seeks to integrate these solutions, instead of trying to solve our problems piecemeal—which only leads to more intractable problems. The book doesn't claim to have all the answers, by any stretch. But it does gather some of the world's clearest thinking about them, as a start. It aims to help build the discussion about survival we have to have.

And it offers a practical way forward, one that can engage every member of our species.

I acknowledge a special debt to the ideas and inspiration of the late Professor Tony McMichael, Professor Paul Ehrlich, Professor Hugh Possingham, Professor Terry Hughes, Professor Will Steffen, Professor Ravi Naidu, Professor Ming Hung Wong, Professor Alon Tal, Professor Bob Douglas, Bill McKibben, Major-General Michael Jeffery, Professor Clive Hamilton, Professor Ian Lowe, Gerda Verburg, Sherestha Saini, Dr. Sean Coffey, Dr. Dennis Hussey, Ian Dunlop, Bill D'Arcy, Peter Day, Dr Mark Stafford-Smith, Dr. Alex Ritchie, Dr. Adrian Gibbs, Dr. Ian Chambers, Richard Heinberg, Professor EO Wilson and David Suzuki.

Canberra, ACT, Australia   Julian Cribb

Contents

1

The Self-Worshipper (*Homo suilaudans*)

Homo sapiens, creatorum operum perfectissiumum. Ultimum & summum in Telluris cortice. (Wise man, most perfect of the Creator's works, ultimate and highest on the surface of the World.)

—*Linnaeus, 1758*

Velvet night enfolds the African savannah. The last light of day vanished half an hour ago; beneath a panoply of starlight filtered by scudding cloud a boy picks his way home across the veldt, following a familiar track. As he approaches Black Hill, the place where his family takes shelter at dusk each day, the ground becomes rough and uneven, limestone boulders litter the grassy slopes leading up to a rocky outcrop, groined by eons of rain and wind into a natural fortress of low cliffs, meandering crevices, shallow caves and shelters—a place even ferocious predators avoid once darkness has fallen.

His attention focused on the uneven footing, the youngster fails to detect the deeper shadow in the tree that overhangs this narrow part of the track, beaten by many feet over long ages. Indeed, the tree itself is hardly to be seen a black silhouette against the fitful starshine. In daylight the tree appears old, rotten, stripped of foliage, devoid of any place to hide, a gaunt object not to be feared. Only on a moonless night like this is it a menace, as the old woman has often warned. But the youth is daring, lithe and strong. Home is close. The path underfoot skirts the boulders, weaving among the rising outcrops—all other ways are far more difficult, treacherous, and just as risky. He should not have stayed out so long, hunting, to prove his prowess and pride. As he passes beneath the outstretched arm of the tree, a shadow blacker than the blackness overhead launches itself silently, blotting out the

© Springer International Publishing Switzerland 2017
J. Cribb, *Surviving the 21st Century*, DOI 10.1007/978-3-319-41270-2_1

dim vault of the sky. The youth knows a moment's panic, total terror and excruciating agony before his neck is expertly snapped. Teeth like daggers sink into his face and skull and, in a series of brutal tugs, the limp body is withdrawn silently into the dry grasses, heaved down the hillside to a lair where a hungry brood awaits.

Huddled safe in their rocky hilltop haven the family wait in vain, wait for the return of day, mourning yet another member of their clan to fall to the ruthless serial killer who stalks them in dreams as well as in reality. Not the first, by any means. One of a long, long line of child victims stretching back tens of thousands of years, hundreds of thousands, millions even.

The killing is real.

It took place sometime between 1.8 and 1.5 million years ago. The victim was a child, probably a young male from a small family group of *Paranthropus robustus*, a ruggedly-built extinct relative of modern humans, who regularly made use of the rocky shelters around Swartkrans, in the bushveld not far from Johannesburg in South Africa. We know how it happened because archaeologist Robert Brain, whose team unearthed the grisly forensic evidence, says

> Another insight into this came to light at Swartkrans when we found that the back of the skull of a child had two small round holes in its parietal bones. I noticed that the distance between these holes was matched very closely by that of the lower canines of a fossil leopard from the same part of the cave. My interpretation was that the child had been killed by a leopard, probably by the usual neck-bite, and then picked up with the lower canines in the back of the head and the upper ones in the child's face. It was then carried into the lower parts of the cave, and consumed there (Brain 2009).

Brain's analysis of the remains of other prey animals, especially baboons, revealed that leopards had a habit of chewing the bones but leaving the hard dome of the skull untouched, a grim testimony of ancient slaughter to a modern humanity which has mostly long since forgotten what it is to be the hunted, rather than the hunter.

Yet, around the same time, and in exactly the same place, another equally remarkable event is taking place: people are discovering the use of fire. And of something far, far more important.

The site of Swartkrans, not far from Pretoria and Johannesburg in South Africa, offers the first definitive evidence for the control of fire by pre-humans. In a memorable letter to Nature in December 1988, Brain and colleagues reported "During recent excavations of hominid-bearing breccias in the Swartkrans cave altered bones were recovered from Member 3 (about 1.0–1.5 Myr BP) which seemed to have been burnt. We examined the histology and chemistry of these specimens and found that they had been heated to a range of temperatures consistent with that occurring in campfires. The presence of these burnt bones, together with their distribution in the cave, is the earliest direct evidence for use of fire by hominids in the fossil record" (Brain and Sillent 1988).

The dating is imprecise, but the rock stratum containing burnt bones and other traces of fire is between a million and a million-and-a-half years old. The child's punctured skull was found in a layer at the same site dated at one and a half million years, or a bit older.

Although there is no direct link between the actual killing and the use of fire, other than the shared location, the inference that fire was first adopted by humans as a defence against predators such as leopards is reasonable and has been widely accepted by archaeologists. It is probable that cooking followed soon after, bringing many health and dietary benefits. All animals are afraid of fire, especially of the vast wildfires that rage across the world's savannahs, fuelled by grasses cured to tinder in summer heat, ignited by lightning strikes, and fanned by hot, fierce winds. Even when these fires die down, animals avoid the smouldering areas. There must have been a special threat, and a special fear, that drove these prehuman creatures—animals with brains not a great deal larger than a modern chimpanzee's—to beat down their natural instinct to avoid fire at all costs, to gather up the embers, to carry them carefully back to the home site and there conjure the flames forth again.

Pre-humans had been walking the grasslands of Africa for at least six million years before fire came to Swartkrans. They had no doubt fled wildfires many times and seen other animals, leopards included, do the same. To conquer their own fear of fire, and to exploit the leopard's, was a spectacular leap forward into the age of humanity. To do such a thing requires a very special skill: the ability to look into the future, to imagine a possible threat—and to conceive, in the abstract, a way of meeting it. The site of Swartkrans captures both moments in the phenomenal ascendancy of humans. This unexceptional, low, grassy African hill with its rocky crown marks the birthplace of wisdom.

To use fire as a defence against leopards requires the user first of all to imagine the family being attacked by leopards in the future   not hard to do in view of the chewed bones found in the leopard's lair, suggesting this was a not-infrequent event. Most animals can imagine a threat from their predators—and the necessity for avoiding it. But then it requires the ability to envisage something the leopard itself fears more than hunger; to step outside oneself and enter the mind of a leopard. Next it demands the ability to see that if one can vanquish one's own instinctive fear of fire, it may confer a decisive advantage in the unequal contest between carnivore and child. And then the ability to see that fire can be carried, nurtured, built up, sustained, both day and night, in an age when matchsticks lay more than million years into the future. Fire itself demands foresight. It must be fed with dry grass, leaves, twigs, logs, tended incessantly for days, weeks, months maybe—through rain, wind and cold. This is not a task for an individual; it requires an act of mutual understanding and co-operation by the whole family group. It probably entails the organised collection and storage of a stock of dry fuel for the hours of dark and days of rain, when no-one dares to venture beyond the cave. If the fire goes out, as the classic 1911 Belgian novel *Quest for Fire* so lyrically recounts, the fate of the entire family hangs in the balance (Rosny 1911).

There were two pre-humans who frequented the cave at Swartkrans, *Paranthropus robustus* and *Homo ergaster*. Sometime, about a million years ago, the rugby-framed *Paranthropus* disappears from the fossil record, a human cousin now lost in the mists of time. *Ergaster*, many archaeologists believe, went on to become us, a main stem human ancestor. It isn't clear who the fire-user was, or which of the two left stone and bone tools lying around in the cave together with traces of hunting and simple butchery. Probably *Ergaster*, who had the larger brain, though still small in comparison to ours (600–680cc against our 1100–1300cc), but possibly both. However, there is something else, even more important, which is singular about fire.

Fire brings light and warmth when darkness falls. For the first time in six million years of roaming the African grasslands, pre-humans have extended the boundaries of daylight into the evening and night. Instead of huddling together in a defensive heap, children at the back, males and strong females to the front throughout the dark hours, the family sits around the leaping flames, gazing into their hypnotic efflorescence, secure in the knowledge that predators will keep away. They have discovered leisure.

Leisure is liberation from the daily grind of survival, the unending cycle of hunting and gathering. It permits the pursuit of things of the hand and mind. It creates space for communication between individuals and among the whole group, the sharing of precious knowledge, the learning and teaching of skills, the shaping and testing of objects, experiments in cookery, the nurturing of children's eager minds, playtime as practice for real life, the emergence of what we now call 'society'. Especially it gives time for the development of that little bone in our throat which sets us apart from the rest of the animal kingdom, the hyoid, the anchor-point for an increasingly agile tongue. Leisure fosters the sounds which later become speech, the gestures, dance, laughter, chant and song which accentuate and amplify its meaning. And with each new sound, the ability to transmit complex ideas grows—and the brains that create and receive them must grow in size too, to handle the vast multiplication in the connections they have to make to process these big, new ideas.

Fire did not create the human brain—but almost certainly spending a third of the last million years sitting around one in a group hastened its expansion. The lack of fire also helps to explain why other social animals, for all their native intelligence, made no similar ascent.

The singular quality which humans developed from this experience of being preyed upon and discovering a way to prevent it was foresight, the ability to look well into the future, discern a mortal threat, understand and neutralise it through changes in behaviour and often by the use of technology. And technology, even if it is a simple stone, bone or wooden implement, requires the maker to first image in their mind the design and production method, and how it is to be used.

The Nature of Wisdom

Foresight is humanity's ultimate skill. The one that set us on a unique path and underpinned all that has since followed. Survival often demands that we first conquer our own primal fears in order to develop the technology or practice that makes us safe. It is something we have never ceased to do, and it lies at the root of all our science, technology, our buildings, our institutions, our vaccines, armies, sewers, fire brigades, healthcare, agriculture, food packaging, transport, traffic lights, high-viz clothing, first aid, clean water, climate science, environmental protection agencies. Like fire, these are all ways of limiting the future mortal risks that life entails.

Our quintessential wisdom is the wisdom of the survivor.

Wisdom, in Greek mythology, is a goddess named Sophia. It is also personified by Athena, who sprang full-armed for battle from the throbbing head of Zeus, and adopted the owl as her emblem. The Greeks understood that wisdom arises from deep thought. But they chose the wise old owl as its totem, probably because the bird's exceptional eyesight enables it to see far ahead, even in darkness.

A workable contemporary meaning of wisdom is 'the ability to think and act using knowledge, experience, understanding, common sense, and insight' (Collins English Dictionary 2014). These entail the skill of rationally envisioning the future. Another definition is 'the skilful application of knowledge'. Unfortunately, a significant portion of humanity is reluctant to practice this informed foresight, and declines to do so, preferring to cling to the status quo. Like our juvenile *Paranthropus* stumbling through the dark, there are people who never envision a real and present danger until it is too late—and, in the age of democracy, their collective myopia overshadows our species' future, especially when politics grants it excessive influence. However, this absence of vision should not be confused with conservatism—the natural precautionary instinct to stick to things which experience has taught us can be trusted and relied upon. There is a world of difference between being cautious and careful about change—and refusing to change.

Today, we owe much of our superabundance of confidence in human wisdom to an eighteenth century Swedish botanist, Carl Nilsson Linnaeus, named by some as the most influential person in all of history for his tremendous life's work in classifying and explaining the relationships between living things—a system now universally adopted and thus exceeding in global influence the impact of dictators and religious leaders (Naylor 2014). He was the first to apply the term 'wise' to the whole of humanity collectively. In 1758, in the tenth edition of his masterwork *Systema Naturae* he formally named human kind according to the two-word system he employed, as *Homo sapiens*—'wise man' in Latin. What on Earth was Linnaeus thinking? And how has it changed us? Has naming ourselves 'wise', in fact made us overconfident, hubristic—a species that rashly deems itself bulletproof against the mounting challenges that surround it as our numbers and demands on the planet multiply? How will it govern our ultimate fate?

Linnaeus was born at Rashult in Smaland, in rural southern Sweden, in 1707 and received his education at Uppsala University, where he began teaching botany in 1730. As a child, when he was upset, his mother used to give him a flower, which immediately calmed him. An early brush with a poor tutor whom he later described as "better equipped to extinguish a child's talents than develop them" left him with a sour taste for formal education which was soon manifest in his skipping school and roaming the countryside in search of interesting plants. Fortunately, a perceptive headmaster noticed his bent and, rather than repress it, gave him the run of his garden and encouraged his botanical studies by introducing him to a skilled local naturalist and doctor, Johan Rothman, who shared with the bright young spark an exceptional library of rare plant books. His fascination with nature kindled, he continued to neglect his formal studies for the priesthood: his concerned father Nils was horrified to learn his teacher's report—that, in his view, the boy would never make a scholar or a priest. Rothman quickly intervened, proposing the lad would make a better doctor than a cleric; he took him in and began his formal instruction in botany, at the same time introducing him to the early systems for classifying living things which the French Jesuit de Tournefort and others had by that time proposed.

The wayward school dropout thus grew into an acute and painstaking observer of nature, with a compendious knowledge of plants, who soon began to perceive relationships between different types of plants and animals which escaped most people, based on detailed observation of their physical attributes. He studied at Lund University, in Skane, but on the advice of his mentor moved to Uppsala University where he came under the kindly tutelage of Professor Olof Celsius, another keen amateur botanist. Linnaeus's true career began with the publication of his thesis on the sexual reproduction of plants, which soon led to his appointment as a teacher of botany. The young man, just 23 at the time, proved a popular lecturer, sometimes attracting audiences of 300 or more. In his reflective moments, however, he began to find fault with de Tournefort's rather arbitrary system for classifying plants—and decided develop his own, which he based on the number of pistils and stamens, or sexual organs, of each plant.

This fertile moment coincided with a university grant to visit and explore Lapland, in the Swedish far north, for new kinds of plants. Despite the region being frozen for half the year and botanically impoverished as a result, the keen-eyed young scholar managed to identify no fewer than 100 plant species that were entirely unknown to the science of his day. These appeared in Linnaeus' exceptional work *Flora Lapponica,* which describes 534 different plants of the region, presenting them according to his scheme of classification based on their sexual characteristics. But his sharp eye was not limited to plants: riding along one day he passed the jawbone of a horse, lying beside the road. Serendipitously, the thought arose "If I only knew how many teeth and of what kind every animal had, how many teats and where they were placed, I should perhaps be able to work out a perfectly natural system for the arrangement of all quadrupeds." Linnaeus was on the way to a profound interpretation of all life that would change forever human understanding of it, and erect the essential scientific platform that has since enabled brilliant naturalists from Cuvier, Owen and Darwin to Dawkins, E.O. Wilson and Attenborough to explain the natural world, and humanity's place in it, to us.

After publishing his account of the botany of Lappland, Linnaeus set to work on the first draft of his epic work, *Systema Naturae.* This was still in manuscript form when he decided to undertake a degree in medicine at the University of Harderwijk in the Netherlands. There he showed the draft to two eminent scholars, Gronovius and Lawson, who were so impressed by it that they agreed to fund its publication, which took place in 1735. In this book, Linnaeus places humans for the first time among the primates, or great apes, based purely on analysis of the species' physical anatomy. In a separate treatise, *Menniskans cousiner* (Man's Cousins), he explained just how hard it was to determine a physical difference between apes and people—even though he clearly understood that from a moral and religious viewpoint it was easy to distinguish between a person and an animal: "In my laboratory I have to behave like a shoemaker at his last, treating man and his body like a naturalist who cannot distinguish him from the apes otherwise than by the fact that the apes have intervals between the canines and the other teeth." The distinction, or lack of it, inevitably caused a public uproar and he hastily explained: "Man is an animal that the creator has decided to endow with extraordinary intelligence and to recognize him as the chosen one, reserving a nobler existence for him. God even sent his only Son to Earth for man's salvation" (Anon.).

In the first edition of *Systema* Linnaeus made it clear that in his view *Homo* had no anatomical features that set him apart from the other primates—the only thing, he felt, that distinguished the human was encased in the ancient motto of the Delphic Apollo, "Know Thyself" (γνῶθισεαυτόν), which he expressed in Latin as *nosce te ipsum*. Self-awareness and the ability to recognise others as human was, in his view, the signal trait of humanity, the one which the other primates lacked. He wrote to the great naturalist Gmelin, who had accused him of saying that humanity was created in the image of an ape, pleading "And yet man recognizes himself. ... I would ask you and the entire world to show me a generic difference between ape and man that would be consistent with the principles of natural history. I do not know of any." All this took place 100 years or more before Darwin and Huxley unleashed a similar debate in the less intellectually open-minded social climate of the mid-nineteenth century. By 1758, when he had had almost a quarter of a century to ponder the relationship, Linnaeus decided that *Homo* stood in need of a more distinctive descriptor than a mere genus name, so he appended a new word. "*Homo sapiens*, creatorum operum perfectissiumum. ultimum & summum in Telluris cortice. (Wise man, most perfect of the Creator's works, ultimate and highest on the surface of the World.)"

He went on to describe *sapientia* (wisdom) as 'a particle of the divine heaven', explaining that the first step in gaining it is the ability to know oneself. He still classified humans according to their physical attributes among the primates, alongside the apes, lemurs and bats, but made separate by this god-given, high ability to know themselves. The Latin word sapiens has three closely-connected meanings: rational, sane and wise. Which of these, if any, Linnaeus explicitly meant to summon up when he chose the word *sapiens* as our species name is lost in the mist of time.

However, the effect of Linnaeus' classification lives with us to this very day: most humans are pleased to consider themselves as wise, or as members of a wise species. Our phenomenal technical achievements of the nineteenth, twentieth and twenty-first centuries have confirmed us in this good opinion of ourselves—one that perhaps confuses mere knowledge and technical ability with true wisdom. Indeed, any attempt to assert that humanity is not wise frequently evokes the same sort of outrage and abuse that the original claim of our descent from a common apelike ancestor produced.

Did Linnaeus, unwittingly, set a terrible trap for humans? Did he, by his simple choice of a word breed into our kind a dangerous over-confidence, complacency and overweening self-satisfaction, a sense that we alone in all creation are intelligent and that the laws which appear to govern all other animals on Earth do not in fact apply to us? Did he blur the boundary between knowledge and wisdom?

Linnaeus lived in an age when, even with the intellectual tolerance of The Enlightenment, it was inadvisable to deny the divine. The last of the great witch trials in Scandinavia took place a mere 14 years before he was born. Just 18 years before this, in 1675, in the parish of Torsacker, 71 people were beheaded and burned for witchcraft while their families and neighbours looked on, reportedly without emotion. The horrors of the 30 Years War between Catholic and protestant Europe, among the beastliest of religious conflicts in all history, were as present to Linnaeus's generation as World War 1 is to ours. So it is not altogether surprising that he chose to explain man as a primate, based on his physical attributes, but cautiously endowed him with a spark of the divine to distinguish him as a species—a concession to the religious fanaticism that could blaze up without warning even in his enlightened times. This was at a moment in history when European thought was precariously poised between blind faith in divinity and predestination—and a growing awakening to the realisation we are in control of, and responsible for, our own destiny and can exercise complete free will. Cleverly, Linnaeus managed to span the theological and philosophical divide by incorporating both meanings into his *sapientia,* by asserting that self-knowledge was god-given. The repercussions of this linguistic ploy have been profound, and accompany the human self-image, narrative and self-regard to this very day, making it harder for us to admit fallibility and error and to correct them. A mere word, and a Latin one at that, might thus sabotage the very attribute that has ensured our survival and ascent so far.

A name is who you are. How humans regard themselves may well hold the key to the fate of our civilisation, and possibly even our species, in the twenty-first century. Wisdom, not knowledge or technology alone, will decide whether we survive and prosper collectively, whether a few survive after some frightful struggle—or whether we all go down in darkness, another evolutionary dead end like *Paranthropus,* lacking the foresight to avoid our own, self-ordained, fate.

Unwise Man

In the science of taxonomy or description and classification of living creatures which Linnaeus has bequeathed to us there is a charming rule known as the 'old fool' principle, summed up as 'the oldest fool is always right'. In science, the purpose of taxonomy is to standardise and stabilise the names of plants, animals and living organisms and so avoid the sort of chaos and confusion which, for example, consumers regularly encounter in the fish market when they find common fish names misapplied by canny traders, trying to palm off a cheap fish as something a bit more expensive. The 'old fool' principle makes it very hard for scientists to change the name of a species chosen by its original namer, unless there is a strongly persuasive reason to do so. This is also known as the principle of priority. It has ensured that Linnaeus' original name for our species has survived virtually unchallenged for more than two and a half centuries.

Modern scientific taxonomy is administered under the International Code on Zoological Nomenclature (ICZN). Without going to its technical details, the Code's rules do allow for species to be renamed provided certain conditions apply: "The valid name of a taxon is the oldest available name applied to it, unless that name has been invalidated." Possible grounds for invalidation include:

• The discovery of new scientific attributes of the named species
• Changes in the common understanding of the species
• Changes found in its phylogeny, or descent
• Correction of an error in its original name
• Lack of a type specimen (known as a holotype) (Segers 2009).

This book puts forward evidence in support of the argument that our species, *Homo sapiens sapiens,* should be urgently renamed, basically, on all five of these grounds.

And that the reason for so doing is now a matter of life and death for hundreds of millions, possibly billions—not just a matter of scientific nicety.

Like our own personal name, our species name directly influences who we think we are, how we see ourselves and our traits, the stories we tell about ourselves and hence, what ultimately becomes of us. It can save us—or condemn us.

For there is one additional ground not embraced by the rules of the ICZN—and it is this: that by insisting of referring to ourselves as 'wise, wise man' we possibly risk our own extinction, and certainly undreamed-of suffering and hardship, due to an operating self-delusion. As a species which deems itself wise, the evidence is now amassing that humanity is collectively behaving no more wisely than a drunken adolescent at the wheel of a very fast and powerful car—ignoring threats to our own life and the lives of others and persisting in the behaviours that most imperil them. We have lost, abandoned, forgotten or diluted the signal quality that set us apart from and above all other species in the Earth over the past one million years: the ability to wisely envision the future, understand it and take well-considered precautions against a bad outcome.

It has become clear that one of the greatest obstacles to wise collective action lies in our mutual self-admiration, our complacency, our conceit—and in the illusion of immunity which they seem to confer when most people think about their future. Too often this attitude is captured in statements like "I don't want to hear any more bad news", "We will solve all our problems with technology" or "God will save us". The first is the cry of those who wish to block out the risks which living entails: however, being deaf to bad news does not abolish it; it simply renders people unprepared. Someone who does not want to hear about risks is acting contrary to the million-year old practice which has guaranteed human survival so far, the practice that gave us fire, and most other technologies since. They are ignoring Darwin's dictum about their own fitness to survive. The second attitude that "We can fix anything" represents a gullible technophilia, or worship of technology, that is over-optimistically blind to its wider or downstream consequences. If we humans have ascended to our present levels of success, health and prosperity through technology then, equally, most of the major threats and perils that now surround us are the result of our misuse, overuse or abuse of those self-same technologies. The unarguable lesson of experience is that each new technology brings with it its own unique set of problems, which in some cases may accumulate to the point where they threaten our continuance as a society or even a species. The future challenge is to design technologies and systems that do not pose such a risk—and where potential downsides are carefully anticipated and avoided in advance. The third argument—God will save us—is simply an abrogation of personal responsibility for one's own, and one's children's fate, and as such unlikely to please any deity.

The proposition that *Homo sapiens* should be renamed is based on the following grounds:

* Any reasonable scientific assessment of the species' current behaviour could not, in the present circumstances be described as 'wise'. The word *sapiens* is therefore a misnomer.

- A common understanding of humanity in the twenty-first century is far removed from the common understanding of the eighteenth century. The name is therefore an anachronism, misleading and no longer appropriate.
- There has been a huge increase in scientific knowledge about human ancestry (phylogeny) since Linnaeus's day, including the essential understanding that several related species of humans have become extinct, and that no form of human is immune to this possibility
- The original name was a poor choice, as even in the C18th it would be hard to describe the bulk of humanity as 'wise', even in respect of other animals, and this is less the case today, as we shall see.
- There is no type specimen (or holotype) of humans, making us the exception among species—although Linnaeus himself has been proposed, and other individuals suggested or have volunteered themselves. The name therefore fails one of the primary requirements of the laws of taxonomy.

The principal reason for discussing a new name, however, is that our present self-lauding title inclines us to a foolish overestimate of our ability to foresee and avoid major risks arising from our own actions, and to a widely-shared blindness to significant existential threats.

Time is running out to re-think who we are. As Ian Chambers and John Humble put it in their thoughtful book *Plan for the Planet*, "We are at a profound point in human history. Never before... have the stakes been so high. The future of the planet and human civilisation as we know it is at stake— and the solutions must be found by this current generation" (Chambers and Humble 2012).

The following chapters explore the latest scientific evidence for each of the ten main existential threats to the human future, its causes—and what can be done to overcome or alleviate it, both at species and individual level. This book certainly does not claim to have all the answers—only that many answers exist and that humanity in the twenty-first century now faces the greatest test of our collective wisdom in our million-year ascent. Whether we succeed or fail is entirely in our own hands, minds and hearts, not just as individuals or nations, but as a species.

And it is now, not in a generation's time, that the decision to survive or fail must be taken.

2

The Terminator *(Homo exterminans)*

The Lord Krishna said:
Doom am I, full-ripe, dealing death to the worlds, intent on devouring
mankind

—*Bhagavad Gita, translation by Mahatma Ghandi.*

When Lonesome George passed, alone and in the dead of night, the world Pinta Island tortoise population slipped quietly into oblivion. George was a member of a subspecies, *Chelonoidis nigra abingdonii*, that had for two or three million years inhabited a single island in the Galapagos. Discovered by Hungarian biologist József Vágvölgyi in 1971, George lived out his last years in captivity and solitude, the only one of his tribe left alive. Desperate efforts by conservationists from the Charles Darwin Foundation to mate him with female tortoises of different subspecies from neighbouring Isabella and Española islands came to nothing. Eggs were laid, but no baby tortoises ever hatched. On June 24, 2012, George was found dead by his carer of 40 years, Eduardo Llerena. It appears he died of heart failure aged around 100 years— not great for one of these large, long-lived creatures, which are known to attain 175 or more (Galapagos Conservancy 2014).

Perhaps it was simply a broken heart. The humans immediately began to fight over who should have custody of his carcass. A Yale University team asserted that the Pinta tortoises still existed in certain genes discovered in another subspecies on another island (Ingber 2012), but if your immediate family is dead and your distant cousins in another land are alive, it doesn't

© Springer International Publishing Switzerland 2017
J. Cribb, *Surviving the 21st Century*, DOI 10.1007/978-3-319-41270-2_2

seem quite the same, somehow. In the final washup, George was simply the latest victim of *Homo sapiens*, whose whalers and fishers used for two centuries to stock up on fresh tortoises to eke out their shipboard rations when they called at the Galapagos and whose feral goats—introduced for similar reasons—stripped Pinta Island bare of all the vegetation the tortoises needed to subsist. Nothing we did subsequently could save him.

Around the world, the mournful tale of Lonesome George is echoing, time and again, with dismaying frequency. On the same day he passed, as many as a 100 other animals and plants blinked into nothingness, unhymned by the world media, blithely unnoticed by the vast majority of *Homo sapiens,* the species that has now come to dominate all others and to occupy the lion's share of the planet's resources.

Like death, extinction—let it be said—is a part of life. Nearly all the plants, animals and organisms that ever lived on Earth are now extinct (Raup 1986). Without extinction, you don't get evolution, adaptation or major change. New species, such as ourselves, cannot emerge to replace old ones, try out new physical and mental adaptations, explore new niches: without it we'd still all be dwellers in the primordial ooze. Extinction has been going on constantly ever since life began here, 3.8 billion years ago. It isn't extinction that is the big worry: it's the *rate* of extinction. To illustrate with a simple analogy, if you smash your car into a tree at a speed of 5 km an hour, you'll probably walk away with scarcely a bruise. Try it at 160 kmh and the effect is catastrophic, for driver and passengers alike. In extinction as well as motoring, speed counts.

Establishing how many species there are on Earth is a task fraught with difficulty: early estimates ranged wildly, from three million to upward of 100 million. However, as techniques improved a team led by Camilo Mora and Boris Worm of Dalhousie University in Canada, modelled species relationships to predict there are probably about 8,700,000 different species currently living on Earth—but possibly as few as 7.4m or as many as 10m (Mora et al. 2011). These include 7.77 million species of animals, 298,000 plants, 611,000 species of fungi, 36,400 species of single-celled animals and 27,500 species of algae. But here comes the crunch: "In spite of 250 years of taxonomic classification and over 1.2 million species already catalogued in a central database, our results suggest that some 86% of existing species on Earth and 91% of species in the ocean still await description," they said. Thus, we're losing living things we don't even know are there. We are losing, forever, a large part of our planet without ever having explored it.

A second vexed question is how fast do species vanish—and is the current rate normal or not? To answer this Juuriaan de Vos of Brown University and colleagues analysed the family trees of living organisms to arrive at an estimate that under Earth's 'normal' conditions—i.e. without human interference or asteroids smashing into it—around one species (out of the roughly 9m species now living) goes naturally extinct every year (De Vos et al. 2014). However, investigations by Gerardo Ceballos, Paul R. Ehrlich and colleagues found "an exceptionally rapid loss of biodiversity over the last few centuries, indicating that a sixth mass extinction is already under way." They concluded that the average rate of vertebrate species loss over the past century has been 100 times higher than the background rate. "The evidence is incontrovertible that recent extinction rates are unprecedented in human history and highly unusual in Earth's history. Our analysis emphasizes that our global society has started to destroy species of other organisms at an accelerating rate, initiating a mass extinction episode unparalleled for 65 million years. If the currently elevated extinction pace is allowed to continue, humans will soon (in as little as three human lifetimes) be deprived of many biodiversity benefits. On human time scales, this loss would be effectively permanent." Averting a dramatic decay of biodiversity and the subsequent loss of ecosystem services was still possible through intensified conservation efforts, the scientists felt—but that window of opportunity was rapidly closing (Ceballos et al. 2016).

The International Union for the Conservation of Nature (IUCN) is the world's oldest and largest global environmental organisation, with more than 11,000 volunteer scientists worldwide supplying it with information on the state of life on Earth (IUCN 2016). Though it works on many aspects of conservation, the IUCN is best known for its celebrated 'Red List' of threatened species, the world's most comprehensive database on the status of imperilled animal, fungi and plant species and their connection to human livelihoods— a list it has maintained and developed for over half a century (The IUCN Redlist of Threatened Species 2016). As of 2016, the Red List had scientifically assessed 83,000 species, finding around 24,000—almost one in three— to be at risk of extinction. It has a publicly accessible internet search engine which enables anyone who is interested, with a little persistence, to discover the current status of any listed life form.

Here is a bird's eye view of what's happening to the world's wildlife, summarised by the California-based Centre for Biological Diversity:

- FROGS: about 2100 of the world's 6300 known frogs, toads and salamanders are in danger, with an extinction rate 25–45,000 times above 'normal'.
- BIRDS: 12% of the world's 10,000 known bird species are classified as endangered, and 200 of these are on the brink of extinction.
- FISH: Worldwide 1851 species of fish—21% of all fish species evaluated— were deemed at risk of extinction by the IUCN in 2010.

- INSECTS: Of the 1.3 million known insect and invertebrate species, the IUCN has evaluated about 10,000 species: about 30% of these are deemed at risk of extinction.
- MAMMALS: Half the globe's 5491 known mammals are declining in population and a fifth are clearly at risk of disappearing forever. 1131 mammals across the globe are classified as endangered, threatened, or vulnerable.
- PLANTS: Of the world's 300,000 known species of plants, the IUCN has evaluated 13,000 and found more than two thirds of these are threatened with extinction.
- REPTILES: Globally, 21% (or 600) of the total evaluated reptiles in the world are deemed endangered or vulnerable to extinction (Centre for Biodiversity 2016).

In 2014 the scientific journal *Nature* estimated that 765 species had vanished since the year 1500, and 5522 were on the brink (Monastersky 2014). On average, it found, between 10 and 700 species were disappearing each week, due to habitat loss and degradation (44%), overexploitation (37%), climate change (7%), invasive species (5%), pollution (4%) and disease (2%). Figure 2.1 shows the most threatened mammal species, by country.

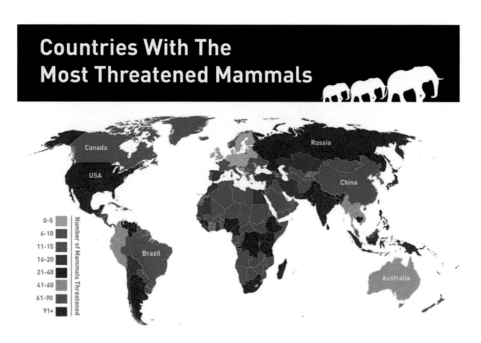

Fig. 2.1 Countries with the most threatened mammals. *Source:* www.theecoexperts. co.uk, data from the World Bank

Mass die-offs of wildlife also appear to be on the rise, according to a study by US researchers. These are staggering events in which more than 90 % of a particular population—often over a billion individuals—perishes. They analysed 727 such events recorded worldwide since 1940, affecting 2407 animal populations, found that their magnitude has increased and was intensifying for birds, fish and marine invertebrates, decreasing for reptiles and amphibians and staying the same for mammals. Most appeared to be caused by starvation, disease, poisoning and multiple stresses (Fey et al. 2014).

Even more alarmingly, a report from the Worldwide Fund for Nature (WWF) found that more than half (52 %) of the world's wild animals vanished in the 40 years post 1970. Land animal numbers were down 39 %, freshwater animals by 76 % and sea creatures by 39 %. The decline in animal, fish and bird numbers was calculated by analysing 10,000 different animal populations representing 3430 different species. WWF said the declines were variously due to over-exploitation by humans (37 %), habitat decline and loss (44 %), climate change (7 %) and other factors (11 %) (see Fig. 2.2) (WWF 2014).

As for animals, so too for the world's plants. A study for the International Union for the Conservation of Nature (IUCN) concluded in 2015 that "More than 20% of plant species assessed are threatened with extinction... the habitat with the most threatened species is overwhelmingly tropical rain forest, where the greatest threat to plants is anthropogenic habitat conversion,

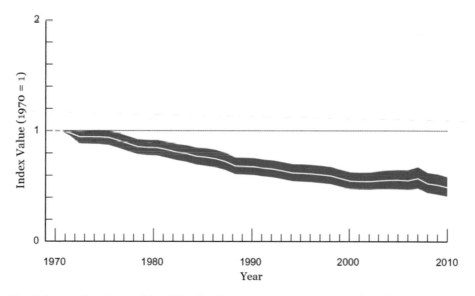

Fig. 2.2 Decline in world wild animal numbers. *Source:* WWF (2014)

for arable and livestock agriculture, and harvesting of natural resources…. Urgent action is needed if we are to avoid losing one in five plant species," it concluded (Brummitt et al. 2015).

Harvard scientist EO Wilson, regarded by many as the world's greatest living biologist, says:

> We are tearing down the biosphere. Without abatement, the current rate of human activity will result in as many as half the species of plants and animals being extinct or on the brink of extinction by the end of the century. I don't think the world can sustain this. It really will be forever (Glancy 2014).

The Frogs: A Modern Tragedy

When Aristophanes wrote his comic play *The Frogs* around the fifth century BC, the little creatures were abundant, their chorus a tuneful backdrop to daily life almost everywhere. By the early twenty-first century frogs and other amphibians, for more than 300 million years among the hardiest and most resilient of lifeforms ever to colonise the Planet, have become the most imperilled group of animals on Earth, their song increasingly stilled. Two in every five of their known species face extinction and there has been a collapse in numbers in all countries and environments, even in the remote wilderness.

Frogs are the perfect illustration of the complex, multi-tyned character of the human assault on the natural world. Their decline and demise is due to no one factor, but to the interaction of many: habitat destruction, infectious diseases, pollution and use of pesticides, climate change, invasive species, and over-harvesting for the pet and food trades.

One of the biggest frog-killers, the chytrid fungus, illustrates the point. This infection has been found in 287 species of frogs and toads, in 37 countries on six continents in a pandemic that is now considered impossible to eradicate (Kriger and Hero 2009). While superficially the infection appears natural, in fact scientists now consider, it was we who spread the chytrid fungus worldwide. Its earliest origins have been traced to South Africa in 1938 and a native frog, *Xenopus laevis*—the African clawed frog—which carries but is immune to it. Around the same time, in 1934, medical scientists made a marvellous discovery that was to change world history for people and for frogs alike: if you inject a frog with a woman's urine, the hormones reveal whether or not she is pregnant. "Soon after discovery of the pregnancy assay for humans in 1934, enormous quantities of the [clawed frog] species were caught in the wild in southern Africa and exported around the world" (Weldon et al. 2004). Following this, the clawed frog became one of the world's lab animals of choice, because it was so easy to keep and breed. Inevitably, some escaped from captivity and established new populations in the wild, alongside native frogs who were susceptible to the fungus and whom they infected, with deadly consequences.

Frog populations anywhere close to modern agriculture, transport or cities were already reeling from a multiple toxic assault from pesticides, air and water pollution, plastic particles, oil and chemical spills, endocrine disrupting chemicals, lead, mercury and other heavy metals. This undermined their health, and their immunity to new diseases. The same poison flood also kills many insects on which frogs rely for their food. Development has led to the draining of marshes, wetlands and shallow aquifers, the emptying of creeks, the clearing and drying-out of once well-watered landscapes to raise grain and cities. And the irresistible decline in frog numbers in turn has spilled into declines in birds, fish, reptiles and small animals that feed on them.

The tragedy of the frogs is a story not only of importance to the fate of life on Earth in general: it is linked to the human destiny also. For we, too, are an involuntary target for many of the same assaults to which we have subjected frogs—and the evidence is mounting that they are starting to exact a heavy existential toll on us too. Frogs, in short, are the canary in the planet's coal mine, their fading calls the early warning of a gross ecological breakdown which will strike humanity harder than anything in our experience. To this we can either choose to succumb—or together take urgent steps to avoid it.

The Hand of *Homo*

That humans are implicated in the dramatic acceleration in loss of species now being seen around the world is no longer doubted by the tens of thousands of researchers who study the issue. "In the past 500 years, humans have triggered a wave of extinction, threat, and local population declines that may be comparable in both rate and magnitude with the five previous mass extinctions of Earth's history," state Rodolfo Dirzo and colleagues in an article in the journal *Science* (Dirzo et al. 2014).

> "We live amid a global wave of anthropogenically driven biodiversity loss: species and population extirpations and, critically, declines in local species abundance. Particularly, human impacts on animal biodiversity are an under-recognized form of global environmental change," they add. "Of a conservatively estimated 5 million to 9 million animal species on the planet, we are likely losing 11,000 to 58,000 species annually".

It is not simply the loss of species like the Pinta tortoise or Africa's northern white rhino which concerns biologists: more alarming still, in their view, is the collapse in abundance of creatures and plants which only a few decades or even years ago, were plentiful. In the case of insects, for example, Dirzo's team found two thirds of the species studies show an average decline of 45 % in abundance. Using information from the IUCN's Red List they found that 60 % of beetle species studied had suffered serious loss in numbers, as had 45 % of ants, 25 % of butterflies—and every single species of grasshoppers and crickets they looked at.

While many people will be little moved by the loss of 'creepy crawlies', such a massive extirpation of insects spills over to affect the numbers of birds, frogs, reptiles and fish that rely on insects as food, and the decline of these in turn affects larger animals. It impairs the successful pollination of plants which provide up to a third of the world's food supply, as well as the renewal of landscapes and forests. Modern plants have evolved largely to depend on insects to fertilise them: lose insects and the whole web of life attenuates and, in some cases, collapses. Like a string of tumbling dominoes, the fall of ecosystems in turn reaches all the way to humans, undermining our own wellbeing through the loss of the services which natural systems provide—clean water, air, food, waste recycling, pollination of crops and seed dispersal of plants, building and furnishing materials, medical drugs, health and recreation.

As the young Australian conservationist Bindi Irwin, daughter of the world-renowned TV naturalist Steve Irwin, graphically explains: "If you keep on pulling one brick after another out of your house, eventually the house falls down" (Surviving Earth 2014). That, say the scientists, is what is now happening, at a planetary scale. This event is so profound it has earned its own name in the geological history of the Earth: the 'Anthropocene defaunation' or, more colloquially, The Sixth Extinction (see for example, Leakey 1996; Kolbert 2014).

Silent Oceans

A major extinction event driven by humans is poised to occur in the world's oceans, similar to the one which has already taken place among land animals over recent history. That's the finding of a study by American marine biologists, who say:

> Humans have profoundly decreased the abundance of both large (e.g., whales) and small (e.g., anchovies) marine fauna. Such declines can generate waves of ecological change that travel both up and down marine food webs and can alter

ocean ecosystem functioning. Human harvesters have also been a major force of evolutionary change in the oceans and have reshaped the genetic structure of marine animal populations. Climate change threatens to accelerate marine defaunation over the next century (McCaulay et al. 2015).

Today's rates of marine extinction "may be the prelude to a major extinction pulse, similar to that observed on land during the industrial revolution, as the footprint of human ocean use widens," they warn, adding that "habitat destruction is likely to become an increasingly dominant threat to ocean wildlife over the next 150 years". However, they consider there is time for humanity to act meaningfully to prevent a wipe-out in the oceans comparable to that taking place on land.

The grim outlook was borne out in 2015 by the Worldwide Fund for Nature (WWF) whose *Living Blue Planet Report,* found "The LPI [Living Planet Index] for marine populations, compiled for this report, shows a decline of 49 per cent between 1970 and 2012. This is based on trends in 5829 populations of 1234 mammal, bird, reptile and fish species" (Worldwide Fund for Nature (WWF) 2015). Looking specifically at tuna and mackerel, the study noted a 74% collapse in numbers and "no sign of overall recovery at a global level". That humans could eliminate almost half of all large sea life across the world's oceans—which span 71% of the surface of the planet—in just 42 years, offers a frightening insight into our destructive capability as a species.

A specific example of how far afield the human hand now reaches is the loss of an estimated 70% of the world's seabird population—equivalent to 230 million birds—since 1950, as revealed in a study by the University of British Columbia. By far the greatest declines were observed in far-ranging ocean-going species like albatrosses (Paleczny et al. 2015). "Seabirds are particularly good indicators of the health of the oceans," lead author Michelle Paleczny commented. "When we see this magnitude of seabird decline, we can see there is something wrong with marine ecosystems. It gives us an idea of the overall impact we're having." A study by Australia's CSIRO and Imperial College London found that 90% of seabirds had fragments of plastic in their gut in 2015 and by 2050 this would apply to 99% of the world's remaining seabird population (Wilcox et al. 2015).

The plight of the oceans is nowhere better illustrated than in the case of Australia's Great Barrier Reef (GBR), the largest living organism on the planet, covering a third of a million square kilometres. In 30 years, half the Reef has died (Australian Institute of Marine Science 2012). In 2016, an estimated 93% of the remaining reef was hit by the worst episode of coral bleaching ever recorded. Some scientists warn that the GBR and most of the world's corals may be gone by 2050 (Koronowski 2016)—the result of a combined assault from human activities including global warming, ocean acidification, nutrient, sediment, oil, chemical and pesticide runoff, dredging, overfishing, boat damage, and plagues of coral diseases, weeds and pests like the Crown of Thorns starfish linked to these human-induced stressors.

If humans can kill off a living organism as large as the Great Barrier Reef through neglect, mismanagement and ignorance, they can kill off anything on Earth, including themselves.

Mass Extinction

The fossil record reveals at least five mass wipe-outs since complex multi-cellular life first appeared in the primeval seas some 700 million years ago, and about a hundred lesser ones. A mass extinction is one in which around three quarters of all the species alive at the time die out. The 'Big Five' are:

- Ordovician-Silurian: a double event about 450–440 million years ago killed off 27% of all families alive at the time, 57% of all genera and 60–70% of known species.
- Devonian: about 350 million years ago, a prolonged event lasting up to 20 million years, eliminated about 19% of all families, 50% of all genera and 70% of all species.
- Permian: 250 million years ago the worst event in the known story of life on Earth took out 96% of all marine species, including all the trilobites. Corals reefs vanished for at least 15 million years. 70% of land species were also lost, for a grand total of 90% of all life on Earth at the time.
- Triassic-Jurassic: about 200 million years ago another vast upheaval eliminated 70–75% of land and water species.
- Sixty-six million years ago the KT or Cretaceous-Palaeogene event, now widely attributed to an asteroid striking the Earth, wiped out three quarters of all known species, including most of the dinosaurs—with the exception of a handful of smaller ones which evolved into today's birds.

The chain of events leading to these five extinctions, especially the older ones, is still debated. The earliest, for example, has been speculatively linked to an exploding star—a supernova—which went off with a bang at the time, and may have been close enough to deluge the Earth with deadly gamma radiation. Or it may have had another cause entirely, which is lost in time. There are at least 15 major theories of natural mass extinction which include:

vast outbreaks of volcanic activity which poison the air and waters; sudden episodes of global warming or global cooling which throw nature and food chains into chaos; rapid sea-level falls bringing death to shallow-water animals; meteorite or comet impacts throwing up vast clouds of dust and causing a freezing 'nuclear winter' in which plants die and food chains collapse; the explosive release of frozen methane deposits in the seabed, disrupting the climate, poisoning the seas and collapsing food chains; loss of oxygen in the oceans and shallow waters caused by the release of nutrients and huge bacterial blooms; 'ocean overturn' where changes in the salt balance and temperature cause the oceans to physically flip upside down, asphyxiating sea life and causing climate havoc. In the worst extinction episodes, it is probable that several of these factors operated in concert with one another to produce the comprehensive tragedies that are graven into the fossil record. It is also likely that extreme climatic change featured in all of them.

The Permian event, for example is variously considered to have started with an outbreak of volcanic flood basalts in Siberia, a massive comet impact or an explosion of methane from the seabed. These in turn released vast volumes of toxic gases and dust into both the atmosphere and waters of the planet. The release of massive amounts of carbon dioxide cooked off by volcanic heat from existing coalbeds and the burning of forests—and recorded in rock strata from that time—may have precipitated a sudden spike in the planet's temperature causing climate chaos and acidifying the oceans, with a resulting collapse in marine food chains. The sudden die-off by a large part of the Earth's life and the erosion of denuded landscapes then poured nutrients into the oceans and fresh waters prompting vast blooms of fungi and bacteria which feasted on the rotting detritus and stripped the waters of their life-giving oxygen, killing off fish and other survivors. For a while, fungi ruled the Earth. Though it is hard to decipher, the fossil record suggests that mass extinctions seldom occur all at once, but instead proceed in a series of distinct pulses, probably due to the sort of cascade of lesser catastrophes described, each one wiping out a new class of animals and plants that had managed to outlive the preceding onslaught. That, researchers fear, is exactly what we are witnessing today—but the causes, and the consequences, lie much closer to home (Ward 2007) .

The Age of Homo

Today, we humans and all living things, inhabit a new age—The Anthropocene. Originally coined to describe our own geological era, the name was adapted by Nobel laureate and atmospheric chemist Paul Crutzen to mean the age in which humans have emerged as a force of nature, with an almost tectonic influence on the planet and all that it contains.

"Human activities are exerting increasing impacts on the environment on all scales, in many ways outcompeting natural processes," he wrote. "This includes the manufacturing of hazardous chemical compounds which are not produced by nature, such as for instance the chlorofluorocarbon gases which are responsible for the "ozone hole". Because human activities have also grown to become significant geological forces, for instance through land use changes, deforestation and fossil fuel burning, it is justified to assign the term "anthropocene" to the current geological epoch. This epoch may be defined to have started about two centuries ago, coinciding with James Watt's design of the steam engine in 1784" (Crutzen 2006).

Others agree about the start date, but link it to the agricultural revolution which in turn launched the human population boom, widespread deforestation of the planet and the universal loss of soils.

The chief fingerprint of the Anthropocene is the gas, carbon dioxide. Released by the burning of coal and oil and the clearing of land, at the beginning of the modern age its level in the Earth's atmosphere was about 270 parts per million (ppm). By 1950, with industrialisation, this had climbed to 310 ppm (Steffen et al. 2007). Today it stands at over 400 ppm and is on track to reach 600 ppm by mid-century. According to the Australian National University's Will Steffen and Paul Crutzen "Since (1950) the human enterprise has experienced a remarkable explosion, the Great Acceleration, with significant consequences for Earth System functioning. Atmospheric CO2 concentration has risen from 310 to 380 ppm since 1950, with about half of the total rise since the preindustrial era occurring in just the last 30 years. The Great Acceleration is reaching criticality. Whatever unfolds, the next few decades will surely be a tipping point in the evolution of the Anthropocene."

When scientists talk about a 'tipping point' they mean a moment when a system suddenly flips from one, relatively stable, state into another. A river once clear, clean and full of life, turns foul, turbid and lifeless due to pollution and sediment. An area of sea once abundant with fishes, crustaceans and shellfish turns dead and sterile due to a massive influx of fertiliser, chemicals and soil. A forest or grassland becomes a desert, from land-clearing, fire or overgrazing. A lake or forest dies from acid rain. A coral reef is overgrown with weed, and is abandoned by its brightly-hued and diverse fishes. A rainforest is cut down, its soils become so acidic that trees can no longer grow and the land is covered in poor-quality grasses. In all these cases it is extremely difficult, if not impossible, to restore the environment to the state that prevailed before disaster struck, at least on human time frames. Too many important species have been lost. Too much has changed, chemically, hydrologically and in terms of the microbial populations that support life. A 'tipping point' is a euphemism: in practicality, it means a point of no return.

Most reasonable people would be appalled if they knew the full extent of the damage they do to their planet and all life on it, simply by the innocent acts of feeding their family and making a home. It is the nature of the modern world that we are separated and insulated from the actual destruction by long industrial and commercial chains that blind us to the realities of mass consumption. For example, environmentalists argue that the extraction of groundwater water by lithium miners has been a major factor in the collapse of flamingo populations on the salt lakes of the Atacama—yet such is the length of the market chain that few owners of a mobile phone, tablet, laptop computer, drill, electric car or other battery-powered device feel personally responsible for the destruction of wild birds, in a desert far, far away (Fischer 2015). And yet they are. Caring citizens of modern society often express deep concern over the extermination of iconic African and Asian elephants and rhinos by poachers, or the impending loss of polar bears from Arctic ice melt—but also seldom feel responsible for the vanishing of fertile soils, the stripping of forests and vegetation and the microbial life that supports them, benign insects, and hundreds of small birds, frogs, or native rodents. Yet, through the global economy we are, all of us, now engaged in this self-harming act of pulling down our own house. Every dollar we spend on food and consumer goods sends out a tiny monetary signal that drives the relentless diminution, destruction and poisoning of forests, savannahs, soils, rivers, oceans, species and clean air. We are the ones whose insatiable appetite for minerals, meat, grain, timber, chemicals and fossil energy is transfiguring our world beyond recognition—and for all time.

The 'Superpredator'

Humans have been implicated in the loss of species for thousands of years. In his celebrated book *The Future Eaters*, palaeontologist Tim Flannery argued that humans were a factor in the extinction of the ice-age megafauna—the giant mammoths, woolly rhinos, cave bears, elks, sloths, giant birds and diprotodons—in the continents of Europe, America, Asia and especially, Australasia (Flannery 2002). Rapid climate change from the frozen world of the last ice-age to the warm world of the Holocene may have pushed these animals to the brink, but there is now little doubt that humans helped shove many of them over it, usually by hunting but more recently by land clearing for agriculture and cities. In Australia, Flannery argues, it was not so much hunting as firestick farming by Aboriginal Australians across the continent that modified the grazing environment so profoundly that the giant kangaroos and diprotodons could no longer survive in it, a continental-scale event that echoes the micro-tragedy of Lonesome George and the Pinta Island tortoises. Today such continental catastrophes are being repeated in Africa and Asia, last homes of the world's megafauna—the elephants, giraffes, lions, tigers, antelope, buffalo, apes and rhinos—which are going down to overwhelming human need, greed and pressures, just as the wolves, lions, bison, bears, wild horses, lynx and aurochs of Europe did in previous centuries.

The bloody thumbprints of humans are to be found all over the crime scene in the disappearance of several hundred large animals during the past 10,000 years[1]. In the last 500 years there is little forensic doubt about who took out the dodo. Or the New Zealand moa. Or the Steller's Sea Cow. Or the North American passenger pigeon. Or the quagga, the sea mink, the Labrador duck, the Great Auk, the Hokkaido wolf, the Tasmanian 'tiger', the Atlas bear, the Texas wolf, the Japanese river otter, the Caspian tiger, the eastern cougar, the western Black rhinoceros, the Formosan clouded leopard. But less well-understood are the means. In a worldwide study of predation, scientists at Victoria University, British Columbia, concluded:

> Our global survey … revealed that humans kill adult prey, the reproductive capital of populations, at much higher median rates than other predators (up to 14 times higher), with particularly intense exploitation of terrestrial carnivores and fishes. Given this competitive dominance … humans function as an unsustainable "super predator," which—unless additionally constrained by managers—will continue to alter ecological and evolutionary processes globally (Darimont et al. 2015) .

Despite our predatory primacy, not every wild animal to become extinct is hunted to its doom. Many, such as small birds, marsupials and plants, have been hardly hunted or harvested at all. Today the main drivers of extinction are more subtle and begin with the overwhelming pressure of the growing human population, our insatiable hunger for resources and the flood of poisons we release when we access them (Chap. 6). Though it is hard for us to conceive, we humans are now so numerous and demanding, that we occupy around 25 % of the Earth's net primary productivity, the organic carbon that is the basis of all life on the planet (Haberl et al. 2007). In other words we commandeer a quarter

---

[1] (For a partial list see http://en.wikipedia.org/wiki/Timeline_of_extinctions).

of all the available energy for life on Earth, rendering that quarter unavailable for other species. But as our population climbs by another third to 10 or 11 billion, as it likely to do this century if current trends persist and, more importantly, if many of those billions attain the affluent lifestyles of America, Europe or Australasia which involves tripling their demand for resources, then by the late century humans alone will dominate well over half of the planet's total carrying capacity. Species that can adapt to us—dogs, cats, cattle, rats, pigeons, roses, corn, cockroaches, ebola, flu, TB, HIV and zika—may prosper and claim their share of the spoils. Others will fade away. Furthermore, besides absorbing the planet's primary energy flows, humans also perturb the Earth system in other ways, through pollution, chaotic change in food webs, landscapes, water supplies, climate and ecosystem services, on all of which other species depend for their survival. A disturbing study by Tim Newbold and colleagues found that, across almost two thirds of the Earth's land surface, species richness has fallen below 10%, considered one of the safe limits for human survival (Newbold 2016). The rise in human populations and extinctions is compared in Fig. 2.3, from the US Geological Survey (Scott 2008).

Of all the human impacts which affect other creatures and plants, by far the largest is our practice of modifying natural landscapes and seascapes, so they support less and less wildlife. The main reason we modify these environments

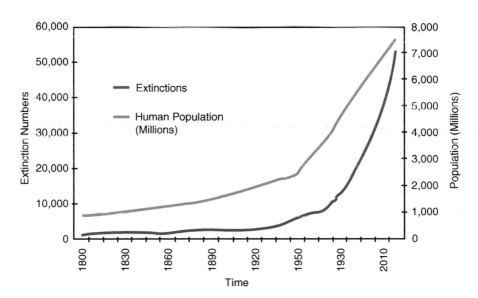

Fig. 2.3 Species extinction and human population. *Source:* US Geological Survey, 2008

is for farming, fishing and grazing in order to supply the food we need each day: from an extinction point of view the human jawbone is by far the most destructive implement on Earth today, and becoming deadlier with each passing day as an additional 200,000 of us sit down to dinner, and call for richer foods (Chap. 7).

Wild species are lost when forests fall, savannahs and rangelands are stripped bare, seas trawled empty, lakes, rivers and aquifers are drained and deserts spread. This is well documented. But there are second-round effects that are equally damaging—torrents of lost topsoil that render waters stagnant or uninhabitable; floods, droughts and wildfires that result from the way we are modifying the landscape and climate; a global outpouring of toxic chemicals into air, water, soil and the food chain, some of which poison animals outright, others of which subtly damage their health or impair their ability to reproduce; the creeping acidification of oceans and lakes. While no single factor may drive a species into extinction, their formidable combination is proving increasingly lethal to more and more life. Together, they render our world less and less habitable for wildlife. And ultimately, for humans too.

Could We Become Extinct?

Over the 3.8 billion-year history of life on Earth, around 99.9 % of all species ever to emerge have gone extinct. Despite this ominous statistic, however, many people nowadays cherish a notion that, somehow, this essential biological truth doesn't apply to us. That we're different. Some imagine our recent technological prowess exempts us from this immutable rule of life, others that a benign deity will intervene to save us. In a good many cases extinction is simply too distasteful or depressing a topic for people to contemplate, and they prefer to stick their head in the sand and pretend it won't happen. This, alas, is not a tactic for survival.

According to the fossil record, the typical Earth species survives for about ten million years before succumbing to its fate or evolving into something else—though there are whole families of long-distance champions like sharks (who have hung round in one form or another for 420 million years), jellyfish (550my) and algae (2+bny). In evolutionary terms this means the modern human race is barely out of the starting blocks compared with these venerable competitors, although our ancestral primate line reaches back 55 million years (Perkins 2013). Our direct lineage extends for only about 4–5 million years, our actual species *Homo sapiens* about 200,000 years and our modern subspecies, *H.s.sapiens* a mere 40,000 years. Until recently the greatest risks of humans becoming extinct came from the natural world but now, as Britain's

Astronomer Royal, Martin Rees points out: "This is the first century in the world's history when the biggest threat is from humanity" (Coughlan 2013).

Lurking somewhere in our gloomy ancestral closet may be a number of hideous crimes against our own kind. The human lineage, back to the time when we shared a common ancestor with chimpanzees, contains at least a dozen different species of human-like creatures, one or more of which are almost certainly our direct ancestors. How many of these distinct 'hominin' species existed rather depends on which palaeontologist you speak to—the issue is debated with ritual calumny in academic circles. On the one hand, the Georgian scientist Dr David Lordkipanidze argues on the basis of a handful of skulls that we are all from a common, but physically highly variable, lineage directly descended from early *Homo erectus* around 1.75 million years ago (Lordkipanidze et al. 2013), while his critics dispute this (Schwartz 2000), arguing instead for anywhere from nine to seventeen distinct species in the family tree of humans—*Australopithecus africanus, A. robustus, Homo ergaster, Homo habilis, H. erectus, H. pekinensis, H. heidelbergensis, H. soloensis, H. floresiensis and Homo sapiens neanderthalis,* to name some of the most prominent (Curnoe 2013). Exactly what became of each of these early human 'cousins'— whether they evolved into us, died out, interbred with or were wiped out by one another along the way—is not known. The most celebrated family mystery is the fate of the Neanderthals: whether or not they died out as a result of genocidal pressure from Cro-magnons—as argued by Jared Diamond (Diamond 1993)—suffered from a change in climate for which they were ill-adapted, were out-competed and starved as a result of colliding with a more advanced hunting culture, or simply interbred with the Cro-magnons and became us. Improved carbon dating indicates they vanished quite suddenly, in less than 2000 years, about 40,000 years ago (Higham et al. 2014), while the discovery of a 50,000-year-old toe bone from the Altai mountains in Siberia has yielded enough Neanderthal DNA for scientists to pronounce with confidence there is a fair bit of it still walking around today, in us, and that a certain amount of interbreeding must have taken place (Prufer et al. 2014). However, the stark and unambiguous message of our own lineage is that *no kind of human is exempt from extinction*, no matter how smart it may deem itself. This is a piece of wisdom we need to ponder as we contemplate and plan our longer-term survival.

History, from the conquests of the New World and India, the rape of Africa, the Mongol invasions, the massacres of the American Plains Indians, Australian Aborigines and Russian Kulaks to the Holocaust and failed Nazi attempt to clear 'Lebensraum' for German settlers in Eastern Europe, makes it abundantly clear that we humans are intensely competitive when it comes to grabbing the resources for living that we covet. Unlike other predatory animals, we have little compunction about systematically exterminating whole races and cultures who stand in our way—a practice still going on today as contemporary urban/agricultural society continues to overwhelm, engulf, digest and eliminate the hunter-gatherer cultures (i.e. those with the greatest knowledge of how to live in balance with nature) of most continents. An unpalatable piece of self-knowledge if we are to survive the next 100 years is that 'genocidal' is inscribed on the human CV from our early days, and is not merely a phenomenon of recent centuries or the province of particular races, creeds and nations. There is a dark tendency to our nature which we must vanquish if we are to avoid being the authors of our own undoing in the twenty-first century.

Possible human extinction through our own actions is now regarded as sufficiently credible a risk to command serious academic attention. In 2004 Britain's Astronomer Royal, Professor Martin Rees, published *Our Final Century* in which he argued humanity has only a 50:50 chance of seeing out this century, based on the dangers of technology run amok (Rees 2004). Professor Nick Bostrom of Oxford University's Future of Humanity Institute (FHI) says:

> Our species is introducing entirely new kinds of existential risk — threats we have no track record of surviving. Our longevity as a species therefore offers no strong prior grounds for confident optimism. Consideration of specific existential-risk scenarios bears out the suspicion that the great bulk of existential risk in the foreseeable future consists of anthropogenic existential risks — that is, those arising from human activity. In particular, most of the biggest existential risks seem to be linked to potential future technological breakthroughs that may radically expand our ability to manipulate the external world or our own biology.

On the upside, Bostrom adds that "Public awareness of the global impacts of human activities appears to be increasing… Problems such as climate change, cross-border terrorism, and international financial crises direct attention to global interdependency and threats to the global system. The idea of risk in general seems to have risen in prominence. Given these advances in knowledge, methods, and attitudes, the conditions for securing for existential risks the scrutiny they deserve are unprecedentedly propitious. Opportunities for action may also proliferate" (Bostrom 2013).

Current scenarios for human extinction (or partial wipe-out) being explored by the Oxford Future of Humanity Institute and others include:

- Severe climate change (+3–6 °C), collapsing world food supplies and ecosystems leading to mass migration, resource wars (Dyer 2009) and disease pandemics (McMichael 2012) (This book, Chaps. 4, 5, 7 and 9).

- Uncontrollable or 'runaway' climate change (+8–30 °C), causing the Earth to overheat to temperatures where it becomes physically uninhabitable by humans or any other large animals (Hansen et al. 2013) (Chap. 5).
- Nuclear wars, arising out of religious, resources, ethnic or political disputes, followed by a 'nuclear winter' of collapsing social order, widespread famine and disease (Chap. 4).
- Developments in information technology reaching a point where human intelligence is exceeded and then supplanted by machine intelligence, a theory popularised by physicist Stephen Hawking (Cellan-Jones 2014) (Chap. 8).
- Chain consequences flowing from research into synthetic biology, nano-technology or quantum physics, such as the unintentional creation of destructive self-replicating organisms, machines or substances, or the breaching of unknown physical boundaries (Chap. 8).
- A global pandemic caused by a newly-evolved or man-made infectious virus, such as a strain of influenza which attacks the brain and spine. These already exist in birds and could cross into humans (Chap. 8).
- Ecosystem collapse; or a more subtle and protracted process in which the progressive decline of climatic, biological and environmental services and scarcity of key resources interacts with loss of intelligence and health as a result of pandemic self-poisoning with man-made chemicals and new diseases (Chaps. 2, 3, 6 and 8).
- A process in which delusion becomes so paramount in politics, business, economics, religious beliefs, popular narratives and the behaviour of society that it paralyses our ability to take effective pragmatic action to save ourselves (Chap. 9).
- More optimistically, that our species successfully evolves from our present form into a wiser type of human with the pan-species ability to communicate, co-operate, nurture, conserve and share wisdom universally—rather than one that prefers competition, exploitation, killing and destruction (Chap. 10).
- An unavoidable Earth system catastrophe such as an asteroid impact or large scale outbreak of volcanism such as may have caused the Permian and/or KT extinctions, or a gamma ray burst from a nearby exploding star.

It will be apparent on reading this short list that most forms of human extinction are avoidable, except, possibly, for the last. However, everything depends upon the degree of wisdom we can bring to bear collectively in anticipating and preventing them from reaching a critical pitch. That humanity has already entered the extinction danger zone is attested in research by some of the world's leading thinkers on this issue—Johan Rockstrom, Will Steffen, Brian Walker, Hans Joachim Schellnhuber and Terry Hughes among others—who identify seven planetary boundaries which humanity ought not to cross for its own safety (and three of which we have already transgressed) (Rockström et al. 2009). We will return to this idea in the concluding chapter.

The take-home message from *this* chapter is this: *extinction is optional*. At least in the current century. It's a choice, for you and me. And our properly understanding the most likely causes is the start of a species-wide process for avoiding them.

Equally, ignoring the possibility of human extinction is a good way to guarantee it. Human survival in the C21st depends less on the malign intentions of the few, than on a majority of good people doing little or nothing to ensure it.

The avoidance of human extinction will demand co-operation across a very much-enlarged species of 10 or 11 billion individuals, on a hitherto undreamed-of scale. It will require the collective wisdom—not just the individual intelligence—to foresee, understand and counteract the self-imposed dangers that confront us. It will necessitate fundamental changes in human nature, belief systems, power-sharing, equality and behaviours—including, especially, a shift from competitive to collaborative thinking (Chap. 10). The main risks and their possible remedies will be explored in coming chapters.

Furthermore, the global decision to avoid extinction has to happen *fast*. As Ian Chambers puts it "It is impossible to overemphasise the urgency with which the human race needs to respond to and manage these global challenges. Time is not on our side. What we do or do not do in the next decade will shape the long-term future of our planet and all who live upon it" (Chambers and Humble 2012).

Finally, it should be observed that, under most of the scenarios described above, complete human extirpation is still unlikely in the current century. A more imminent risk is the collapse of civilisation in the chaos arising out of the uncontrolled burgeoning of several dangers and their interaction with one another, rather than any single cause. This is the complexity crisis, which humans will undoubtedly face in the twenty-first century.

A Plague of Teddy Bears

In recent decades the number of toy stuffed animals in the world has multiplied faster, far faster, than humans. Poor households now harbour several of these cute effigies of the natural world, while rich homes boast dozens and occasionally hundreds. Of the 70 or so new toys that the average American child is presented with each year (Tuttle 2012), a tenth or more are toy animals. Some malls and shopping centres have retail outlets whose sole trade is in stuffed toys. Museums and even conservation bodies flog millions of replica wild animals, soon to vanish from the Earth. If a Martian statistician were to conduct a global census, they would probably find that stuffed toys now outnumber humans and other animals several fold, and might even conclude they are the true masters of the Earth…

It may seem whimsical to use something as apparently innocuous as teddy bears as an indicator for man-made eco-collapse and extinction, but they are emblematic of how detached humans have become both from the natural world and from the realities of survival in it. There is something unhinged, disconnected and rather pathetic about a creature which devotes so much time, energy and money to rubbing out real, live animals—and replacing them with lifeless surrogates, mostly made from the very petrochemicals that contribute to extinction. On the one hand our love of stuffed toys betokens our sentimental attachment to aspects of the disappearing natural world. On the other it bespeaks our brute indifference to the fate of actual wild animals with whom we share this world: we can love something fake, artificial, childish and anthropomorphic, but care little for the real thing. While first and foremost a failure of empathy and 'humanity', our love affair with stuffed toys also symbolises a profound failure of wisdom, the inability to grasp that we run the risk of 'stuffing' ourselves, along with the rest of Nature in our thrall.

Preventing Extinction

It is entirely feasible for humanity to slow and maybe even to halt the rising tide of extinctions—but not by the half-measures now in place around the world. Zoos, national parks, marine reserves, cryogenic 'arks', conservation and breeding programs, tree-planting schemes and private collectors may salvage a few fragments of the jigsaw of life which we inherited—but not the big picture. These are the product of wise individuals, who understand the scale of the losses, and are doing their very best to staunch them within the limits of insufficient resources and funding and a largely uncaring, unwise society. Only a wise human *species* can halt the current losses (see below and Chap. 7).

For every person working to conserve an animal, a plant, a landscape or marine area, a million are spending their money in ways that, often unbeknown to them, guarantee that the destruction will continue and even increase.

What would a wise humanity do? Here are some possible approaches, drawn from a growing scientific consensus, with potential to make an Earth-wide difference. Significantly, all of them contain a 'win-win' for humans as well as for the natural world and improve our prospects of surviving the twenty-first century with our civilisation intact. Probably the most important and far reaching, articulated by biologist E.O. Wilson, is to set aside half the Earth for the rest of life: "The only way to save upward of 90 per cent of the rest of life is to vastly increase the area of refuges, from their current 15 per cent of the land and 3 per cent of the sea to half of the land and half of the sea. That amount, as I and others have shown, can be put together from large and small fragments around the world to remain relatively natural, without removing people living there or changing property rights" (Wilson 2016a).

What We Must Do

1. Replace half the world's farmed and grazed area with sustainable, climate-proof intensive food systems, mainly in cities and in coastal waters. This will enable the progressive 're-wilding' of an area of 25 million square kilometres (equal in size to the continent of North America) and its return to its natural vegetation and wildlife.

   Pathway: see Chap. 7 for more detail but broadly this entails the rapid shift of half or more of the world's food production capacity into cities to protect it from climate shocks and water shortages and to recycle nutrients. This demands radical change in urban planning (to recycle water and nutrient waste), to encourage food production, accelerate investment and increase R&D into greenhouse, hydroponic, aquaponic and bioculture systems.

2. Sustainable grazing (eg 'precision pastoralism') of the world's rangelands will enable livestock numbers to be reduced, carbon to be locked up, vegetation and water cycling to be restored and pastoral incomes improved. This will lead to many wild species being far better conserved across the savannahs, which cover 40% of the Earth's land area (see Chap. 7)

   Pathway: the concept of 'precision pastoralism'—using satellites and automated mustering to balance feed availability with livestock numbers—allows far more sustainable grazing of rangelands and better incomes for pastoralists. It will be up to governments to drive this by ensuring the availability of technology and training.

3. Replace destructive fishing methods with sustainable forms of aquaculture on land and at sea, based on farmed algae.

Pathway: aquaculture is already taking off as ocean fish catches dwindle and the cost of farmed fish production falls with technological advances. It will be greatly accelerated by the farming of algae as a major new feed supply for both farmed fish and other livestock as well as for human food and renewable transport fuel.

4. Replace coal, oil and gas with renewable energy. This will eliminate the world's main source of toxic pollution, which is currently harming wildlife as well as all humanity directly and indirectly, through brain poisoning and reproductive dysfunction, immune breakdown and climate change.

   Pathway: See Chap. 4. Detailed pathways and options for climate change mitigation has been laid out by the IPCC (IPCC 2014a) and in many individual government reports[2]. They include strategies such as accelerated investment in renewables, carbon cap-and-trade systems, distributed energy generation, energy efficiencies in industry, transport and cities, smart energy technologies, reafforestation and revegetation of landscapes and recycling of materials, most of which hold additional benefits for the natural world in terms of reduced toxicity and increased wilderness.

5. Build a global biosecurity network to combat the introduction and impact of invasive species.

   Pathway: under development. Requires far stronger quarantine and species import/export rules as well as control of marine pests, and exotic insect and fungal introductions. Increased priority among government agencies of biosecurity risks.

6. Develop a plan to progressively restore the world's great forests, manage the oceans (especially outside sovereign borders) and cleanse the world's seas, rivers and fresh waters of toxins, plastics and eroded soil.

   Pathway: progress in this regard has been slow in some areas and has stalled in others. It can be reignited by a global 'Clean Up the World' strategy (Chap. 6), and reinvestment in programs such as the UN's REDD reafforestation scheme.

7. Build into all food and consumer goods a small charge to fund the repair or prevention of the ecological damage caused by their production. This should be regarded as a wise re-investment in natural capital, not an 'eco-tax'.

   Pathway: the simplest way to do this is through a consumption tax on food that is earmarked specifically for reinvestment in natural capital and repair of damaged landscapes and waters. To avoid regressivity, the poor can be exempted, supplied with food stamps or other concessions.

---

[2] (See, for example, Germany: http://www.eea.europa.eu/soer/countries/de/climate-change-mitigation-national-responses-germany).

8. Use the funds so raised to pay the world's 1.8 billion farmers and indigenous people to act as on-the-ground stewards of global biodiversity and
fund conservation programs for vital habitats and keystone species.
   Pathway: see Chap. 7

With an issue such as extinction the individual, even if wise enough to
understand and lament it, often feels helpless to prevent it. The good news is
that this need no longer be the case. Here are some measures we can all take
in our own lives to ensure the survival of as many other lifeforms as possible.

What You Can Do

- Be an informed consumer. Learn which foods and goods degrade and
  destroy the natural world and which heal it—and exercise your economic power and freedom to send a clear signal to industry, your nation
  and the world economy. Freedom isn't just a right—it's a responsibility.
- Use the internet and social media to learn the scientific facts of extinction and share them with friends, family and followers. Play your part
  as an educator and leader in the online global conservation movement. Stand up for endangered and 'keystone' species.
- Educate your children about the value of wildlife and natural landscapes, how they support us—and what we lose when we degrade or
  destroy them.
- Support politicians and companies with a track record for devoting
  real resources to protection of wildlife and landscapes.
- Avoid products that use plastics, pesticides, endocrine disruptors,
  VOCs and other contaminants that kill or incapacitate wildlife.
- Choose foods and consumer goods that reduce human pressure on
  the natural environment and encourage 're-wilding'.
- Work through local volunteer, social, religious and sports groups to
  repair your local environment, restore its species and spread the
  word about sustainable consumption.
- Don't buy any more stuffed toys. Spend the same amount on a good conservation body or activity (like tree planting) and save a real animal to
  delight your grandkids. Get them involved in wildlife sponsorship schemes.

3

The Degrader (*Homo eversor*)

Thou madest him to have dominion over the works of thy hands; thou hast put all things under his feet

—*Bible, Psalms 8, verse 6.*

The human story in the twenty-first century will be dominated by a titanic global struggle—economic, political, scientific and military—for resources. On this, to a significant degree, turns the fate of civilisation.

In every prior age till now the bounty of the Earth was ample to sustain the ascent of human society. Scarcities, when they occurred, were local, regional or else the result of human interference or mismanagement. Now, with the advent of the postmodern era, a Rubicon has been crossed: the physical demands of seven to ten billion humans, each aspiring to a higher standard of living, are combining to exceed the Earth's carrying capacity. Put simply, we are using more stuff than the planet can renewably provide.

Human use of natural resources amounts to some 75 billion tonnes per year—or 10 tonnes per annum to support each one of us. That demand has grown tenfold in a century, from seven billion tonnes in 1900, and is due to reach 140 billion tonnes by 2050 (OECD 2015a). These resources fall into two main sorts—'non-renewable' resources like minerals, fossil energy, industrial and construction materials, and the so-called 'renewables' such as agriculture, forestry and fisheries—which are now proving not so renewable after all (Chap. 7). Then there are the key environmental resources of water, land, biodiversity and atmosphere.

© Springer International Publishing Switzerland 2017
J. Cribb, *Surviving the 21st Century*, DOI 10.1007/978-3-319-41270-2_3

To put this in a personal perspective, over his or her lifetime each citizen of Earth will (at our present rate of demand):

- Use 99,720 tonnes of fresh water (Fischetti 2012), two thirds of it in the form of food
- Cause the loss of 750 tonnes of topsoil (Wilkinson and McElroy 2006)
- Consume 720 tonnes of metals, manufacturing and construction materials (World Resources Institute 2015)
- Use 5.4 billion British thermal units (Btus) of (mainly fossil) energy (Energy Information Administration 2015a)
- Cause the release of 288 tonnes of carbon dioxide (World Bank 2015a)
- Cause the release of 320 kilos of industrial chemicals, many of them toxic (Cribb 2014)
- Waste 13.4 tonnes of food (Gustavsson et al. 2011a).

While the sheer size of your personal impact on the planet may come as a surprise, it is modest compared to what may occur as the human population expands to between 10 and 12 billion (Gerland et al. 2014) and as, according to PriceWaterhouseCoopers, the world economy grows "at an average rate of just over 3% per annum from 2011 to 2050, doubling in size by 2032 and nearly doubling again by 2050" (PriceWaterhouseCoopers 2012). The OECD, for example, foresees world demand for metals alone expanding from 5.8 to 11.2 billion tonnes between 2002 and 2020 (OECD 2015b). Paradoxically, the efficiency with which the world uses its increasingly scarce resources is declining, not improving—due chiefly to the rise of the newly-industrialising countries, which use less efficient industrial processes, as the globe's manufacturing powerhouses (Resource Efficiency: Economics and Outlook for Asia and the Pacific (REEO) 2011).

Resource over-use has two dimensions. The first is scarcity, which everybody understands and which the market quite often remedies with a substitute—although, in cases such as water and phosphorus, there are no substitutes. The second is pollution, which is far less easy to remedy and which can result in the degradation and destruction of other vital resources: for example, agricultural overuse can destroy soils, rivers and even seas; fossil fuel overuse can poison the air we breathe, reduce our children's intelligence, cause cancers, destabilise the climate and acidify the oceans.

Our use of resources will more than double in the coming half-century and, since the actual resources themselves physically cannot, this runs the risk of unleashing a series of global 'train-wrecks'. Since humans customarily contest resources with ferocity, the main consequence of this will be a heightened risk of conflict—between nations, religious faiths and sects, industries, corporations and social groups—as well as price volatility, sudden shortages, economic shocks, the corruption of markets and governments, and the poisoning of landscapes and whole generations.

This, self-evidently, is an unwise course for *Homo* to pursue. So, what are the main issues? And what are the wise alternatives?

Dry Times

To support the average citizen of Earth takes around 1386 tonnes of water a year. This is known as our 'water footprint' and consists of all the water used to produce our food, consumer products, or provide the services on which we rely: our indirect use of water is many times larger than our personal use. In total, humanity goes through more than 9 trillion tonnes of fresh water annually. The largest users by volume are the United States and China. However, the average American (2800 t) consumes nearly three times more water than does the average Chinese or Indian (1000 t): this is mainly due to America's meat consumption which accounts for almost a third of US water use (Hoekstra and Mekonnen 2011).

"Freshwater is a scarce resource; its annual availability is limited and demand is growing," warns the Water Footprint Network, a global partnership of governments, universities and water bodies. "The water footprint of humanity has exceeded sustainable levels at several places and is unequally distributed among people. There are many spots in the world where serious water depletion or pollution takes place: rivers running dry, dropping lake and groundwater levels and endangered species because of contaminated water. The water footprint refers to the volumes of water consumption and pollution that are 'behind' your daily consumption" (Water Footprint Network 2015). Of these, food grown by agriculture accounts for the vast bulk—around 70 %, or 970 tonnes a year per person. So just feeding you requires nearly 3 tonnes of water a day.

It is commonly assumed that the world has abundant water for everyone and that rainfall and the natural hydrological cycle will supply the lack, but so rapidly has industrial, agricultural, energy and urban demand burgeoned that this outdated view no longer holds true. Four billion people, worldwide, already face severe water scarcity (Mekonnen and Hoekstra 2016). While the global population has tripled over the past century, our use of water has grown sixfold. In 2015 the World Economic Forum ranked water supply crises as the highest of all global risks in terms of their potential impact on the human future—more harmful even than money crises, the spread of nuclear weapons and our failure to adapt to a changing climate (World Economic Forum 2015).

Water scarcity worldwide is being made worse by the rapid depletion of groundwater resources in virtually every country where well-water is used to grow food. Groundwater is one of the largest resources in the planet, accounting for 95 % of the planet's available fresh water. Furthermore, it supplies a significant portion of the surface water in our dams, rivers and lakes and is responsible for maintaining landscapes. Humanity's total extraction of groundwater is conservatively estimated by UNESCO at 1000 cubic kms (a trillion tonnes) per year, of which about 67 % is used for irrigation, 22 % for domestic purposes and 11 % for industry (Van der Gun 2012). In theory the world has eight to ten million cubic kilometres of groundwater—but much of this is in the wrong places, is inaccessible, saline or too expensive to pump. In places where ground water is relied on to irrigate crops, such as the north of China, the Indo-Gangetic region, the Middle East, North Africa and Midwestern USA, water tables have been falling by a metre or more a year for decades—indicating that the water is being withdrawn much faster than it naturally replenishes, which can sometimes take centuries. In a world where nearly half our food is grown by irrigated crop production, the depletion of these reserves poses a real risk to food security within a generation (see Chap. 7).

At the same time as we are mining the world's groundwater unsustainably, climate change is eating into another vital reserve—the cap of snow and ice atop our high mountain chains that supplies substantial flow to our great rivers, and is known as the 'water tower'. Studies by the World Glacier Monitoring Service, an international scientific collaboration, reveal that ice losses from mountain glaciers in the decade 1996–2005 were twice those of the previous decade—and four times greater than the decade before that. These losses have been growing ever since measurements began in the 1940s and correlate strongly with the rise in global temperatures recorded over the same period (World Glacier Monitoring Service 2015). Glaciologists report "clear evidence that centennial glacier retreat is a global phenomenon" and "the rates of early twenty-first century mass loss are without precedent on a global scale" (Zemp et al. 2015).

Since mountain glacial meltwater provides a significant part of the flow in large rivers like those of the Indo-Gangetic plain, Central Asia and continental North and South America, the shrinking, and in some regions complete loss of, mountain glaciers poses a growing threat to food production, groundwater recharge and to those cities which rely on it for drinking water.

Beyond this, the world's rivers and lakes are themselves in a parlous state. Fifty thousand large dams fragment our major river systems globally, causing declining water quality, increased sedimentation and pollution (especially from tonnes of mercury now accumulating in their waters), and exacerbation of global warming through the methane emissions that often bubble from their muddy bottoms (International Rivers 2014). The world's 50 largest river basins are sundered by nearly 6000 major dams—703 on the Mississippi, 374 on the Yangtze, 228 on the Parana, 184 on the Danube and 183 on the St Lawrence. Among the great waters most affected by the combination of dams, over-extraction and climate change are the Ganges and Indus, the Amu Darya and Syr Darya in Central Asia, the Colorado and Rio Grande, the Yellow, Amazon, Danube, Mekong, Nile, the Tigris and Euphrates, the Murray-Darling (World Preservation Foundation 2016). The most polluted rivers on Earth are the Hai Ho, Wisla, Dneipr, Tigris-Euphrates, Yellow, Danube and Mississippi. International Rivers, a global research partnership, states: "The evidence of planetary-scale impacts from river change is strong enough to warrant a major international focus", adding "…dams should become an option of last resort for managing water and generating electricity… No more dams should be built on the mainstem of rivers which play a crucial role for the sustainability of freshwater ecosystems."

In 2015 Bolivia's second largest lake, 2700 sq km Lake Poopo, dried up completely and its largest, Titicaca was in trouble, threatening the city of La Paz. In Mongolia, changing rainfall patterns have caused a third of its lakes to dry up (New Scientist 2016). In China, Hubei—'the province of a thousand lakes'—has lost 90% of them (Peryman 2012). In the Arctic, one third of tundra ponds (thaw lakes) have vanished (Andresen and Lougheed 2015). In Uzbekistan, the Aral Sea—the world's fourth largest lake, has mostly dried out (Quobil 2015). In Niger, Lake Chad, once a vast water body supporting 50 million people, is 95% gone (Lakepedia 2015).

"Many of the world's lakes are in crisis," explains Lakenet, a scientific network involving 100 countries. "Diversion of lake water for use in irrigation and industry, invasions of exotic plant and animal species, and contamination by toxic substances and nutrients from industry, farms, municipal sewage, and polluted urban runoff are common on a scale today that significantly threatens lake ecosystems on every continent but Antarctica" (Lakenet 2015). The main problem affecting the world's lakes is eutrophication, caused mainly by fertiliser and soil run-off from farming, which trigger algal blooms and fish kills. "Fertilizer is the biggest single driver of nutrient pollution globally. Worldwide fertilizer use is expected to rise 145% between 1990 and 2050," says Lakenet. Echoing such concerns, the 13th World Lakes Conference warned:

The ecological state of the lakes worldwide has deteriorated alarmingly during the last decades – and the Chinese lakes are a sad example of it. The 24,800 Chinese lakes cover an area of over 80,000 sq. km – and with a few exceptions nearly all lakes are heavily polluted or on the verge of drying up. Everybody knows that something must be done. There is international consensus on the need for integrated lake management and technical approaches, which are already available. …industrial firms, however, ignore environmental rules in force and prefer to pay (low) fines if they are caught discharging waste water into lakes instead of treating the effluents before. …the irrigation of farm land has led to the shrinking of the Aral Sea, once the world's fourth largest water body to 10% of its original size (Global Nature Fund 2008).

In 2014 the world's cities were home to four billion, or 54%, of the Earth's citizens (UN Department of Economic and Social Affairs (DESA) 2014). By mid-century there will be seven billion urbanites, more than three quarters of humanity. Megacities such as Guangzhou-Shenzhen will harbour 120 million inhabitants, and several will exceed 40 million. With the growth in these vast conurbations comes an insatiable thirst—today's cities swallow an estimated 1.5 trillion tonnes of water and this is expected to double to at least three trillion tonnes in line with their populations and with rising living standards. On top of this demand, the world energy sector is also expected to double in size, imposing its own massive demand for, and impact on, dwindling water resources. Coal mines, gas and oil fields all involve the use, disruption, pollution and loss of immense volumes of fresh water—as the increasingly bitter global confrontation between farmers and frackers demonstrates. However, when it comes to water wars, local farmers—and food security—nearly always lose out to multinational energy giants, who can afford to buy more politicians. This clash will, within decades, confront many regions with a stark choice between food or fossil fuel—or between food and city life.

In the Age of 'Peak Water', in which human demands collide head-on with finite supplies, water scarcity may spell war. While the majority of international water disputes are still resolved amicably, the timeline maintained by Peter Gleick of the Pacific Institute reveals both the rising crescendo and tempo of confrontations over water that have taken place over history, and especially in the early twenty-first Century (Gleick 2015).

The water issue has been described first here because, in all probability, it will be the first of the great resource scarcities to strike at our civilisation, arriving over the coming generation, faster even than the full impact of global warming. By the 2030s, according to one UN report, global water demand could exceed supply by as much as 40% (Schuster-Wallace and Sandford 2015a).

Water is also a supreme example of human unwisdom, of our greed, our wastefulness, our disorganisation, our failure to anticipate the future and to take the necessary corrective action. There is more than enough fresh water on this planet to meet our needs and those of all life, but on the whole we manage it very badly, we are ignorant of the extent of the resource and the rates at which it is replenished or recycled, we squander it using old, wasteful technologies, we value it very low, and we dump everything from our excrement to our industrial and military waste to our plastics, fertilisers, drugs and household poisons into it, thereby ensuring it is unfit for use—or at least too costly to cleanse to drinking or food-growing standards.

Many practical solutions to the threat of global water scarcity exist, but human society and its governments remain reluctant to implement them. Among the best of these solutions are: to recycle all urban water; to move food production from high water-use systems to low-use; to accurately quantify water resources as a prerequisite for effective management; to manage the entire water resource, surface and subsurface, within a region as a single unit; to restore the 'small water cycle' of local rainfall by re-greening the landscape (Kravcik et al. 2008); to store more water underground in aquifers instead of in surface dams; and to place a sufficient price on water to send a clear signal to different users to stop wasting and start saving it.[1]

Not Just Dirt

"A rough calculation of current rates of soil degradation suggests we have about 60 years of topsoil left," Sydney University's Professor John Crawford told TIME magazine in an interview. "Some 40% of soil used for agriculture around the world is classed as either degraded or seriously degraded—the latter means that 70% of the topsoil, the layer allowing plants to grow, is gone. Because of various farming methods that strip the soil of carbon and make it less robust as well as weaker in nutrients, soil is being lost at between 10 and 40 times the rate at which it can be naturally replenished" (Crawford 2012). In a separate report, Professor Duncan Cameron of the University of Sheffield's Grantham Centre for Sustainable Futures found that nearly 33% of the world's arable land had been lost to erosion or pollution between 1975 and 2015 (Cameron et al. 2015).

---

[1] This is complicated, but it involves maybe setting price differentials for different water uses, depending on societal priorities.

Soil degradation is among humanity's largest impacts on the planet—and is typically the one that is least understood or of concern to city people, in spite of the real threat it poses to their future. American soil scientists Bruce Wilkinson and Brandon McElroy calculated that we displace around 75 billion tonnes of topsoil worldwide every year, based on contemporary levels of erosion measured in farmers' fields (Wilkinson and McElroy 2007). This is nearly four times greater than the natural erosion which takes place Earth-wide without human agency. Most of this lost soil eventually ends up at the bottom of the ocean, borne on the wind or in rivers, and so can never again be used to grow food or forests. While the lion's share of the erosion is due to land-clearing, crop cultivation or overgrazing, a significant portion is caused by badly-designed urban development, drainage, roads and engineering works. Since soil typically takes from thousands to millions of years to form through natural weathering of rock and the workings of biology, such a colossal loss of the world's topsoil represents an un-staunched haemorrhage of one of the main things which keeps us alive.

The destruction of topsoil has contributed to the collapse of civilisations in the past, notable cases being the Mayans, the Greeks and the Romans. "Many ancient civilizations indirectly mined their soil to fuel their growth…" writes David Montgomery, author of 'Dirt' (Montgomery 2007). "Such problems are not just ancient history. That soil abuse remains a threat to modern society is clear from the plight of environmental refugees driven from the US southern plains Dust Bowl in the 1930s, the African Sahel in the 1970s and across the Amazon Basin today." South African researchers calculate that soil loss has already reduced African food production by 8%, with worse to come (Scholes and Scholes 2013). Scientists and aid agencies increasingly link the onset of the Syrian civil war and refugee crisis to a savage drought which started in 2006 and turned thousands of farmers off their land (UNCCD 2014). UNCCD executive secretary Monique Barbut said: "When land degradation reaches a level at which it seriously threatens people's livelihoods, it can turn into a security issue. This is because land is so closely linked to basic human needs, such as access to food and water. If land degradation interferes with the fulfilment of these needs, it can lead to conflicts over scarce land and water resources, spark food riots or turn smallholder farmers into refugees. It is striking that many of today's violent conflicts are taking place in countries with vulnerable dry ecosystems" (Barbut 2014).

The United National Food and Agriculture Organisation cautions there are many impacts of soil degradation: "Firstly, billions of tons of soil are being physically lost each year through accelerated erosion from the action of water and wind and by undesirable changes in soil structure. Secondly, many soils are being degraded by increases in their salt content, by waterlogging, or by pollution through the indiscriminate application of chemical and industrial wastes.

"Thirdly, many soils are losing the minerals and organic matter that make them fertile, and in most cases, these materials are not being replaced nearly as fast as they are being depleted. Finally, millions of hectares of good farmland are being lost each year to nonfarm purposes; they are being flooded for reservoirs or paved over for highways, airports, and parking lots. The result of all this mismanagement will be less productive agricultural land at a time when world population is growing and expectations are rising among people everywhere for a better life" (UNFAO 2015a).

"Soil erosion is most serious in China, Africa, India and parts of South America. If the food supply goes down, then obviously, the price goes up. The crisis points will hit the poorest countries hardest, in particular those which rely on imports…The capacity of the planet to produce food is already causing conflict," Crawford told TIME. Rich countries are not immune: they will have to cope with floods of fleeing refugees—as in the Syrian crisis, whose deep drivers many now attribute to a combination of desertification and climate change—increasingly volatile food prices, spreading ill-health in their own populations, conflicts arising out of 'food wars' and failed states.

Soil degradation is not an isolated problem. In particular, it causes the degradation of fresh water—turning rivers and lakes turbid and blocking dams with sediment, poisoning them through chemical and nutrient runoff. It causes dust storms, injurious to human health and, when it enters the ocean is chiefly responsible for the lifeless 'dead zones' now spreading along heavily-populated coastlines around the planet and for the loss of coral reefs. Though the links are seldom appreciated by either consumers or farmers, soil erosion thus contributes to the loss of sea fisheries.

There are subtler linkages too: today's crop varieties, according to Crawford, are specially bred to cope with the depleted conditions of the modern soils in which they are farmed. These crops are much lower in micronutrients—and higher in carbohydrates, a major driver of the global obesity pandemic, as well as the other diet-related diseases that together now claim the lives of two thirds of all humans (WHO 2012). So soil loss is, ironically, one of the factors in a more overweight humanity.

In short, global soil degradation is undermining human health and civilisation's long-term prospects for survival, planetwide. There is, as yet, no end in sight. Yet it is a problem easily solved, as we shall see (Chap. 7).

Falling Forests

Of the major resource issues facing the world, forests have received the great-est publicity, due chiefly to the efforts of the conservation movement and their visual appeal to the media.

Forests now cover about 31 % of the Earth's land surface, just under four billion hectares, which is about one third less than in pre-industrial times when they cloaked 5.9 bn/ha. The UN FAO reports that deforestation reached its peak during the 1990s, when each year the world lost on average 16 million hectares of forest. There was then a loss of 129 million hectares between 1990 and 2015, according to the UN FAO's Global Forest Resources Assessment (2015), which found "The bulk of the world's forest is natural forest, amount-ing to 93 percent of global forest area or 3.7 billion ha in 2015. From 2010 to 2015, natural forest decreased by a net 6.6 million ha per year (8.8 million ha of loss and 2.2 million ha of natural forest gain). This is a reduction in net annual natural forest loss from 8.5 million ha per year (1990–2000) to 6.6 million ha per year (2010–2015)" (UNFAO 2015b).

Despite this superficially encouraging situation, the UN Environment Program cautions "Unfortunately, very few countries have any estimates of the(ir) actual rates of deforestation" (UNEP 2007), a diplomatic way of say-ing that many have good reason not to disclose their records. Also, beneath the raw data of areas of forest loss lies a far more complex and disturbing picture of changes in forest density, quality and species mix. Forests are not only being degraded by direct human activity such as logging, road-building and land-clearing for agriculture, but also by human-mediated factors such as pest invasions, climate change, wildfires, species losses, weed invasions, water table changes and especially fragmentation, which jeopardises long-term forest health. Notably too, much of the 'replacement' of old forests which is shoring up the numbers, often consists of monocultures of oil palms or other commer-cial trees. These are frequently introduced from elsewhere and fail to compen-sate for the species richness lost when the original forest was cut down.

In the second decade of the twenty-first century, every major category of forest type on the planet is in decline (Forest health 2015). While the current rate of loss of the world's forest resources does not directly imperil civilisation on its own—unlike scarcity of water and soils—it has a number of adverse effects that transmit global shockwaves. These include:

- Loss of ecosystem services such as clean water, air and breathable oxygen
- Displacement and loss of forest-dependent human cultures and communities

- Loss of soil stability and nutrient recycling, with resulting pollution and degradation of rivers and lakes
- Loss of timber, fuelwood, animal, food and medicinal products with resulting economic impacts on consumers
- Accelerated global warming caused by the release of carbon normally stored by forests
- Changes in local rainfall patterns due to the replacement of forests with hot, bare areas which respire less moisture (the 'small water cycle')
- Accelerated loss of species, especially in the tropics and subtropics
- Increased poverty.

The destruction of the world's forests is primarily due to greed, bad management and corruption on the part of many governments and timber companies, and is thus not easy to remedy. Even countries which encourage tree planting can seldom replace the original richness of the lost forest. However, schemes such as the UN's REDD (Reducing Emissions from Deforestation and Forest Degradation) program, which funds farmers and local people to replant forests as a carbon sink to counter global warming (UN-REDD 2008) offer considerable hope by creating an economic incentive for renewal, Such measures may dilute the overwhelming economic impulse for destruction which has devastated world forests in the past half century—though they are, as yet, poorly supported by the rich nations.

Desert Cancer

"Desertification spreads like cancer; it can't be noticed immediately," explains Wadid Erian, a soil expert with the Arab League. He instances Syria, where drought displaced hundreds of thousands of people, ruining farmers, swelling cities and fuelling the discontent that led to civil war—and Darfur in western Sudan, equally devastated by warfare exacerbated by shortages of water and fertile soil (Discovery Newsletter 2013).

About half the world's land surface consist of 'dry lands', where rainfall is low, soils fragile and evaporation rates are high. Desertification occurs when these arid land are over-cleared, over-cropped or over-grazed, stripping away the vegetation that holds their soils and recycles their local moisture, and when this and climate change alter their rainfall patterns.

"Desertification is a phenomenon that ranks among the greatest environmental challenges of our time," cautions the United Nations, adding regretfully "Yet most people haven't heard of it or don't understand it."

In actuality, desertification is coupled with soil loss and deforestation. It directly impacts on about 1.5 billion people in 168 countries, and leads to the abandonment of about 12 million hectares of productive land which could otherwise have fed about 60 m people (United Nations 2015). Its impact is particularly severe in Africa, China, the Indian subcontinent, Central Asia and the Middle East. It is, says the UN, "…a global issue, with serious implications worldwide for biodiversity, eco-safety, poverty eradication, socio-economic stability and sustainable development." Scientists estimate that fertile land is now turning to desert at rates 30–35 times greater than in past history (UN 2015).

The original Rio Earth Summit in 1992 ranked desertification alongside climate change and extinction as one of the main threats to the sustainable development of human society. Technically, this loss of the world's dry lands is reversible—by tree planting and landscape revegetation, soil stabilisation, improved water management, conservation farming and 'precision pastoralism'. In practice, while contributing to desertification by its economic demands, the wealthy world does not want to pay for reversing it: as a result, far too little is being done, and far too late, to avoid disaster in many regions, which will eventually spill over to affect everyone.

Ocean Mining

Even in the late twentieth century many people believed it inconceivable that human demands could possibly exceed the bounty of the world's vast oceans, or cause them such harm as to undermine their health and deplete the life they hold. This is no longer true.

"Today, 61% of commercially important assessed marine fish stocks worldwide are fully fished, 29% are overfished," stated the UN Food and Agriculture Organisation (UNFAO 2015c). "About 90% of large predatory fish stocks are already depleted. Our oceans and seas are under risk of irreversible damage to habitats, ecological functions, and biodiversity because of overfishing, climate change and ocean acidification, pollution, unsustainable coastal area development and the unwanted impacts from the extraction of non-living ocean resources." Figure 3.1 shows the peak and decline in world fish catches since the mid-1990s.

Around three billion people rely on seafood for much of their diet, and the mayhem described by FAO makes one thing crystal clear: the amount of food taken from the seas by conventional fishing cannot double in line with the anticipated doubling in world demand for food. Far from it.

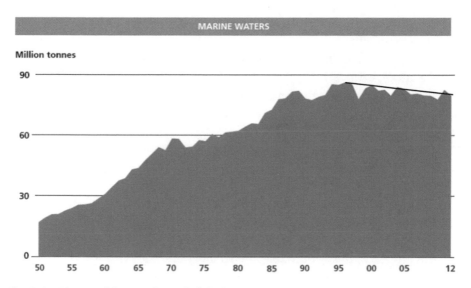

Fig. 3.1 The world passed 'peak fish' in 1994. *Source*: UN FAO, SOFIA, 2014

The damage is exemplified in three issues: dead zones, coastal destruction and acidification.

Dead zones, large areas of ocean, bays and estuaries lacking in oxygen and hence largely lifeless have metastasized around the planet in recent years, driven by a man-made flood of soil, nutrients and toxic chemicals from the land. There are now some 470 dead zones, covering a total area of 245,000 km² (Diaz and Rosenberg 2008).

To this is added the degradation and loss of an estimated third of the world's coastal environments—mangrove forests, seagrass beds, dune systems, bays, estuaries and salt marshes, beaches and coral reefs: these are not only the nurseries for many fish, birds, shrimp, turtles and other marine animals but also hold around half of the ocean's stored carbon, which is now being released into the atmosphere where it accelerates global warming (Conservation International 2015).

"Ocean acidification is sometimes called 'climate change's equally evil twin, and for good reason," says the Smithsonian Institution (Smithsonian Institution 2015). "At least one-quarter of the carbon dioxide (CO_2) released by burning coal, oil and gas doesn't stay in the air, but instead dissolves into the ocean. Since the beginning of the industrial era, the ocean has absorbed some 525 billion tons of CO_2 from the atmosphere, presently around 22 million tons per day." This dissolved CO_2 is changing the chemistry of seawater, turning the oceans slowly but surely more acidic—with inestimable effects on the billions of organisms, from oysters to corals to plankton and diatoms, which rely on alkaline water conditions to form their calcium skeletons and which constitute the foundation of the ocean food chain. Scientists fear just such a process drove other great marine extinctions in the past—and that humans may, inadvertently and in ignorance, have triggered a similar event with potential to strip the oceans of their life, to our own profound loss.

Energy Struggle

Cheap energy is the blood supply of modern civilization. To keep the world ticking over requires the use of about 550,000,000,000,000,000 British thermal units (550 quadrillion Btus) of primary energy each year (US Energy Information Administration 2015b). To illustrate, the typical Canadian consumes around 400 million Btus a year to maintain their lifestyle, the average German 165 m, the average Argentinian 90 m, the average Chinese 80 m and the average Egyptian 42 m. To satisfy this gargantuan global energy hunger in the second decade of the twenty-first century took 33 billion barrels of oil, 120 billion cubic feet of gas, 8.5 billion tonnes of coal and 20 trillion kilowatt hours of electricity every year.

Of this total, fossil fuels supplied around 80 % of all primary energy and renewables about 20 % in the years 2013–2015 (Renewables 2014). See Fig. 3.2.

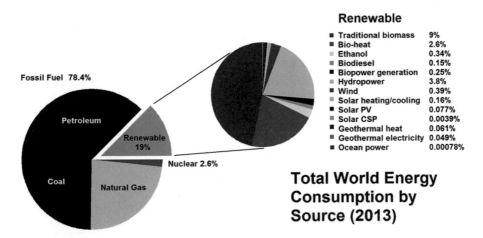

Fig. 3.2 Total world energy consumption, by source. *Source*: REN 21Renewable 2014 Global Status Report

The US Energy Information Agency (US EIA) projects that by 2040, world energy demand will rise to 820 quadrillion Btus, due mainly to economic growth in the newly-industrialised and developing countries. This would entail a 30 % increase in global carbon emissions.

Despite much debate over 'peak oil', the world is not in imminent danger of running out of oil *per se*—or, indeed, of gas and coal—but rather of depleting those reserves which are cheap and easy to exploit. Proven reserves were estimated by the International Energy Agency (IEA) in 2014 to be: 142 years' worth of coal, 64 years of natural gas and 54 years of oil. However, for oil in particular, the world needs to find one new Saudi Arabian-sized oil province every 5 years to keep pace with demand—at a time when world car production has been outpacing world oil production (International Organisation of Motor Vehicle Manufacturers (OICA) 2013). To satisfy human energy demand alternatives will become essential. The IEA cautions that:

- The global energy system is in danger of falling short of the hopes and expectations placed upon it.
- Advances in technology and efficiency give reason for optimism, but sustained political efforts will be essential to change energy trends for the better.
- Global energy demand is set to grow by 37 % by 2040, but the development path for a growing world population and economy is less energy-intensive than it used to be (IEA 2014).

Ian Dunlop, a former senior executive in the international oil, gas and coal industries and latter-day advocate for urgent climate action, points to another serious and widely-unrecognised limitation to the world energy mix: "To get energy out, you have to put energy in. The surplus energy drives our society. However, the ratio of energy-out to energy-in (EROEI) has been declining" (Dunlop 2014). To maintain an industrial civilisation, you need an EROEI of around ten-to-one, Dunlop says. New oil returns roughly 25-to-one. Unconventional sources such as shale oil, tar sands and coal-to-liquids produce far less net energy per unit of effort, and only about half or less of the energy required to sustain the modern economy on which civilisation depends. In short, they are not an answer to world energy scarcity—and nor indeed, by the same measure, are the present crop of renewables, although technological advances may lift their performance.

For fossil fuels, however, the decisive issue of the twenty-first century will not be availability—it will be how society reacts to climate change (Chap. 5) and to the slow poisoning of a generation of children (Chap. 6).

Our Big Feet

The combined effect of humanity's rising demand for resources is described by the Global Footprint Network (GFN). This worldwide academic group has operated since 2003 to document the effects of 'global overshoot', the taking by humans of more materials, food and energy from the Earth than it can supply or renew over the long haul (World Footprint Network 2016). The GFN is famous for its little graph (Fig. 3.3) which depicts the effect of humanity's over-use of our planetary credit card.

"Today humanity uses the equivalent of 1.6 planets to provide the resources we use and absorb our waste. This means it now takes the Earth one year and six months to regenerate what we use in a year. Moderate UN scenarios suggest that if current population and consumption trends continue, by the

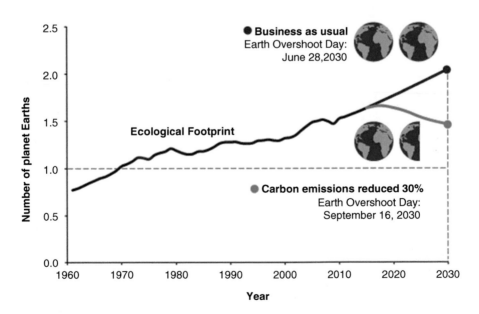

Fig. 3.3 How many Earths does it take to support humanity? *Source*: © 2016 Global Footprint Network. www.footprintnetwork.org

2030s, we will need the equivalent of two Earths to support us. And of course, we only have one," the WFN says.

It goes on to caution that turning resources into waste faster than waste can be turned back into resources puts us into 'ecological overshoot', running down the very things on which all humans and life on Earth depend: "Overshoot also contributes to resource conflicts and wars, mass migrations, famine, disease and other human tragedies—and tends to have a disproportionate impact on the poor," it adds.

It is the rich—countries, cities and individuals—who tread most heavily and heedlessly on the Planet, consuming on average five or six times the amount of resources needed by a poor person. It is they who most endanger the future of civilisation by their insatiable demands—not the far more numerous humans at the lower end of the socioeconomic spectrum. According to the WFN's national comparisons, the typical Saudi or Emirates citizen, for example, devours 11 times more resources than the typical Kenyan or Rwandan. US, Danish and Belgian citizens use eight times as much stuff, Australians and Canadians seven times, Germans, Swiss, Britons and Japanese five times. These, therefore, are the citizens upon whom the greatest burden of responsibility falls to lead the conversion of the global system from one of overuse, waste and pollution to one of saving, recycling and reuse.

Eco-Collapse

The constellation of resource scarcities, extinction and decline in environmental services such as clean air, water, healthy landscapes etc, in the view of many scholars, carries with it potential for a collapse in modern civilisation, if not its actual extinction. This was the central thesis of Jared Diamond's 2005 book *Collapse: how societies choose to fail or survive* (Diamond 2005) in which he identified eight factors in common between modern civilisation and failed civilisations of the past:

1. De-forestation and habitat destruction
2. Soil problems (erosion, salinization, and soil fertility losses)
3. Water management problems
4. Overhunting
5. Overfishing
6. Effects of introduced species on native species
7. Overpopulation
8. Increased per-capita impact of people.

However, Diamond warned, this situation is made worse in the modern context by man-made climate change, the build-up of man-made toxins in the environment, energy shortages and the increasing dominance of human use over the Earth's photosynthetic capacity (i.e. our ill-conceived destruction of the world's forests and grasslands).

Many people have difficulty grasping the idea that humans could so devastate our own environment as to render it unable to support us. One way to envisage this was proposed by John Schramski, a professor at the University of Georgia, whose team has studied the energy balance on Earth represented by its plant life. They concluded that if we continue to destroy plants and trees at present rates, it will imperil our own continued existence: "You can think of the Earth like a battery that has been charged very slowly over billions of years," Schramski says. "The sun's energy is stored in plants and fossil fuels, but humans are draining that energy much faster than it can be replenished." The researchers calculate that 2000 years ago the earth had about 1000 gigatonnes (billion tonnes) of energy stored in the form of plant carbon. Since then human activity has halved this colossal reserve. "If we don't reverse this trend, we'll eventually reach a point where the 'biomass battery' discharges to a level at which Earth can no longer sustain us," he cautions (Schramski et al. 2015).

Writing in *The Bulletin of The Atomic Scientists*, ethicist Phil Torres argues that eco-collapse exists alongside nuclear weapons and global warming as a 'clear and present danger' to the human future. "The repercussions of biodiversity loss are potentially as severe as those anticipated from climate change, or even a nuclear conflict.... Biodiversity loss is a "threat multiplier" that, by pushing societies to the brink of collapse, will exacerbate existing conflicts and introduce entirely new struggles between state and non-state actors. Indeed, it could even fuel the rise of terrorism" (Torres 2016).

In their analysis of contemporary global trends population scholars Paul and Anne Ehrlich comment

… today, for the first time, humanity's global civilization—the worldwide, increasingly interconnected, highly technological society in which we all are to one degree or another, embedded—is threatened with collapse by an array of environmental problems. Humankind finds itself engaged in what Prince Charles described as 'an act of suicide on a grand scale', facing what the UK's Chief Scientific Advisor John Beddington called a 'perfect storm' of environmental problems (Ehrlich and Ehrlich 2013).

The Ehrlichs point out that a collapse would be unavoidable in the event of either a small nuclear war or major famines, but equally could come about through global toxification and the compound failure of key ecosystems on which we rely for survival. They note that scientists have repeatedly warned humanity about all these issues—and repeatedly been ignored. As a result, many governments are still asking questions like 'how do we feed ten billion people?' instead of 'how do we reduce the size of the population to something the Earth can sustain?'

"There is not much evidence of societies mobilizing and making sacrifices to meet gradually worsening conditions that threaten real disaster for future generations. Yet that is exactly the sort of mobilization that we believe is required to avoid a collapse," the Ehrlichs conclude.

The End of Mining?

However, there is hope that the depletion of the Earth's reserves of energy and minerals can be brought to an end. Mining, with agriculture, is one of the cornerstones of civilization. Humans have been extracting minerals from the Earth for tens of thousands of years, first flint for tools and ochre for paint, then clays for pottery and ultimately the metals that gave rise to the bronze, iron and computer ages. Currently we mine about ten billion tonnes of metal ores each year (OECD 2008) and move around 1000 billion tonnes (or more) of rock, soil and water and their chemical constituents to do so (UNEP 2002). So it may come as something of a shock to consider that the twenty-first century could witness the end of mining.

The authors of the demise of one of our most ancient and significant technical arts will not be green protestors and their politicians, as some miners might fancy. They will be women. The women of the world.

Women have already taken the decision—without referring to men—to reduce the world's population. While the population itself continues to grow slowly (due to people living longer as well as to more babies being born) in absolute terms, fertility—the number of babies born to each woman—is dropping in every continent and virtually all nations and societies. This is the result of an unspoken global consensus comprising billions of individual decisions by the female of the species as she acquires education, self-empowerment, income, a career, clean water, healthcare, greater freedom from poverty, violence and oppression and a safer home. According to the UN, the number of babies per woman has fallen from 4.4 in the 1970s to 2.4 today and will approach 2 (or zero population growth) soon after the mid-century. The dramatic implications of this development for the survival of civilisation and the human species will be explored in Chap. 10. For the moment, however, we need to consider what it means for our resource use.

The universal decision by women to reduce the number of their babies, and even to have no babies at all, means that at some moment in the mid-to-late twenty-first century the world will attain "peak people" and then—with luck—commence a slow, steady dwindling in numbers to a more sustainable level. At least, that is the wise scenario: the alternative is considered in Chap. 4.

Peak people means, quite simply, that there will be limited demand for new metals and materials extracted from raw resources, as nearly all the metal humanity will ever need is already available in our waste stream. Instead of having to move and process up to a 300 tonnes of rock, soil, ore, tailings and slag to extract a single tonne or less of pure metal, we can sift the waste stream for whatever we need—iron, copper, aluminium, zinc, tin, manganese, or any of the forty-odd rare metals in your mobile phone—using an array of sophisticated new technologies, such as solar furnaces, currently under development. Indeed, manufacturers of complex products like cell phones and tablet computers will probably design them for easy recycling of their component metals, many of which are rare and expensive, and encourage consumers to swap their old model for a new one: this is known as 'cradle-to-cradle' manufacturing. In a world of sharply declining demand for new raw materials, the actual instruction to 'cease mining' will in fact come from company accountants and shareholders, faced with amassing evidence that it is more profitable to recycle existing metals than to extract them *de novo*: intelligent mining firms will quickly reconfigure themselves as 'resource re-processors'.

The end of mining will also bring with it an end to one of humanity's greatest acts of contamination, one which unleashes around a trillion tonnes of soil, rock, acid, polluted water and toxic metals annually into the environment, to the detriment of rivers, lakes, wildlife, the air, the living environment, food safety and human health. Recycling metals not only makes economic sense; it brings substantial health benefits, increases our prospects for survival and ends the senseless destruction of the natural world.

The 'end of mining' will bring to a close an epoch of human history and technology that has extended for more than 6000 years, replacing the 'extractive era' with the 'age of reuse'. This illustrates the sort of transformation in human behaviour and thought which must occur if we are to learn to live within the finite bounds of our planet—and avoid the kinds of disasters which so many scholars and scientists now foresee.

Avoiding Scarcity

Resource overuse endangers the future of civilisation and humanity at several levels—through pollution and the poisoning of both planet and people, through the degradation of vital natural systems including fresh water, soils, forests, biota, the atmosphere and oceans, through the economic and political instability which scarcity engenders and through the conflicts it ignites.

Thrift—the art of frugality or saving scarce resources—is often viewed from today's prodigal perspective as one of those antique virtues beloved of bygone generations, with declining relevance to the twenty-first century. Not so. It is—and has always been—a skill essential to the survival of human society and civilisation. Your grandmother knew thrift was important to survival, and you'd do well to heed her advice. Past societies that didn't preserve food for winter often perished. Modern societies that ignore their resource limitations equally take a blind gamble on their own future.

Fortunately, there are practical and profitable alternatives to the overuse of resources. The International Resources Panel (IRP) of the UN Environment Program, among others, makes a strong case for 'decoupling' resources from economic growth or 'dematerialising the economy', as it is sometimes known (International Resources Panel 2011). Essentially this means creating an economy in which continued growth is possible without relying entirely on growth in consumption of material resources like energy, minerals, timber, water and so on: it means generating more wealth, or gross domestic product, for fewer resources expended. This relies on achieving greater technical efficiency in resource use, on recycling material resources as close to 100 % as possible, and on using our brains to find better ways to substitute or replace old resources that are becoming scarce or producing unwanted side-effects like pollution and climate change. A classic example of decoupling is the substitution of clean renewable energy, such as wind, tide, solar and bio, for fossil energies like coal, oil and gas (although even these so-called 'clean' energies impose a significant demand for rare minerals and materials). Another is 'industrial ecology' where one industry makes productive use of the 'wastes' generated by another, thereby eliminating waste and turning it into money. A third is 'product stewardship' in which manufacturers harvest the materials they need to make new products from the waste stream of the old ones, creating a cradle-to-cradle lifecycle.

An important example of decoupling is soilless food production. This exploits technologies like hydroponics, aquaponics, and biocultures to grow crops, livestock, fish and protein in ways that do not require soil and which have a very much reduced need for water, artificial fertilisers or pesticides because they recycle both water and nutrients and use very hygienic means of production. This will allow the world to grow as much food as it needs while dramatically reducing the human impact on soils, landscapes, oceans, fresh water and wild species. More on this in Chap. 7. The IRP concludes "efficient technologies do exist for both developing and developed countries to significantly reduce resource intensity and, where feasible, achieve the absolute decoupling of resource use" (International Resources Panel 2014).

At the same time as we reduce our dependence on material resources, we need to grow the part of the economy that relies on things of the human mind—art, science, entertainment, literature, software, IT, design, multimedia, the internet, social media, the caring professions, culture, tourism, services, sport, leisure and so on. This 'knowledge economy' (OECD 1996) or 'creative economy' (Cunningham 2013) is where the jobs of the future will mostly exist. It has a dramatically smaller material footprint than the 'old economy', creates far less pollution or risk of climate change and above all, is bounded in its growth potential only by the limits of the human mind itself. In theory, the possibilities of such an economy are therefore infinite…

What We Must Do

1. Recycle everything that is scarce, especially water, metals, timber, plastics, food, nutrients, textiles and building materials.

    Pathway: the key is to improve the economics and social attractiveness of recycling through a combination of enhanced consumer demand for recycled products and government incentives. This can be achieved through public awareness and education, by industry marketing its successes, through research and development and via the tax system. Approaches such as 'industrial ecology' and 'cradle-to-cradle manufacturing' should be widely encouraged and adopted (Chap. 6)

2. Decouple economic growth from the limitations of material resources by focussing on growth within the 'knowledge economy', and in the products and occupations of the inexhaustible human mind and imagination.

    Pathway: this should be promoted by government and industry as a 'win-win' for the 21st century economy, which can achieve continued economic and jobs growth while reducing our physical demands on the planet.

3. Price natural resources at their true value. Eliminate subsidies.

    Pathway: Governments must be pressured by citizens to eliminate inherent contradictions in their policies, such as subsidising fossil fuels (which encourages their wasteful use) while at the same time trying to mitigate climate change by reducing their use.

4. Phase out the permanent dumping of waste and replace it with waste-stream 'mining'.

    Pathway: Economic signals are essential. Heavier charges, prohibitions and penalties may be needed to discourage waste disposal, dumping and pollution. Monetary and social incentives may be needed to encourage recycling/reuse.

5. Progressively replace fossil fuels with renewable, non-polluting energy— such as solar, wind, tide, geothermal, algal oil, thorium cycle nuclear, for example.

    Pathway: the market is the best tool to achieve this transition rapidly: carbon cap-and-trade, consumer education leading to greater demand for clean energy, disinvestment in fossil fuel companies (stimulating them to switch to renewables), elimination of subsidies, tougher environmental restrictions on mining and extraction, pollution penalties etc.

6. Replace a significant part of agriculture with intensive, sustainable urban food production systems; replace fishing with aquaculture and algae culture:

    Pathway: See Chaps. 2 and 7

7. Strengthen worldwide co-operation on the governance and restoration of fresh waters, forests, soils, landscapes and oceans.

    Pathway: greater national and local collaboration and knowledge sharing via the internet, UN agencies, international agreements and farmer-to-farmer.

8. Provide women with the resources and power to curb human population growth, raise and educate children better.

    Pathway: universal education, healthcare, female empowerment and family planning. This can be delivered collaboratively at global, national and local levels, also via the internet and social media and by responsible transnational corporates.

9. Place a charge on all resource use, including food, and re-invest it in replanting and regenerating forests, grasslands, desert margins, rangelands, fisheries and coastal ecosystems.

    Pathway: this is best achieved via the taxation system or by means similar to carbon trading, which place a higher cost on less efficient and unsustainable production systems and reward the more efficient and those that recycle, regenerate and renew, thereby encouraging the transition. It is essential that farmers receive price signals which facilitate the switch to regenerative agriculture and miners likewise to metal reprocessing.

10. Establish universal, free, education to ensure that each of Earth's citizens understands the need to sustain and care for the vital Earth systems and resources that support them.

   Pathway: already among the Sustainable Development Goals, there is a need for global awareness that education is not just desirable – it will also play a determining role in the human destiny. Free education is not a luxury, but rather a precondition for our surviving the 21st century. Countries with free universal education advance more rapidly.

What You Can Do

- Conscientiously reduce your household consumption of material resources, personal waste and use of energy.
- Learn what goes into the food and goods you purchase and how sustainable it is.
- Choose a diet, activities and consumer goods which sustain, rather than destroy or waste, natural resources, and which heal you and your family rather than cause sickness.
- Participate actively as a global citizen in the universal online consumer and lifestyle movement to share knowledge, ideas and advice for a sustainable future.
- Use your power as a consumer to send a signal to manufacturers, farmers, industry and government that we value sustainable goods and services—and will reward those who deliver them.
- Use your power as a voter to support politicians whose deeds (not just words) demonstrate they are committed to a sustainable knowledge economy.
- Teach your children to prize clean air, water, soil, forests, wildlife and the ocean as much as life and liberty. For that is what they are.
- Live more like your grandparents: reduce possessions, repair and keep things longer, recycle, rediscover the virtue (and satisfaction) of thrift.

4

The Butcher (*Homo carnifex*)

*From the dawn of consciousness until 6 August 1945, man had to live
with the prospect of his death as an individual; since the day when the
first atomic bomb outshone the sun over Hiroshima, mankind as a whole
has had to live with the prospect of its extinction as a species.*
—Arthur Koestler, Janus: A Summing Up, 1978

Twenty five thousand years ago, two battlelines of fierce warriors arrayed
themselves against one another and unleashed a hail of missiles. The boldest of the
leaders, seeking to inspire his followers, sprang forward to charge his foes—only
to be laid low, pierced by their spears and smashed by heavy war boomerangs. An
elegant piece of rock art from the Kimberley region in north Western Australia
depicts the dramatic confrontation between the warriors captured in the act of
discharging their missiles at one another, and the heroic (or dismal) death of one
of their leaders. Experts think the artwork, from an enigmatic phase of Australian
Aboriginal art known as Gwion Gwion, may date from 20 to 25,000 years ago
and be the world's first known battle painting. Though there are only seven warriors on one side and eight on the other, in its depiction of an heroically fallen
war-chief the image evokes the Homeric confrontation of Achilles and Hector in
the Trojan war (c 1250 BC) and innumerable battle stories since. In so doing it
reminds us that the custom of organised warfare is bred into our species, alone,
since the dawn (Australian Rock Art Initiative 2011).

It is also a stark reminder that, in the twenty-first century, the militant spirit
of *H.s.sapiens* is one of the most likely pathways to self-destruction and that
the risks today are, if anything, greater than they have ever been. For example,
world military spending in 2015 was $1.7 trillion (Stockholm International

© Springer International Publishing Switzerland 2017
J. Cribb, *Surviving the 21st Century*, DOI 10.1007/978-3-319-41270-2_4

Peace Research Institute (SIPRI) 2016), compared with global investment in food and agricultural science of around $50 billion. This suggests a contemporary definition of humanity as "a species which spends 34 times more on better ways to kill itself than it does on better ways to feed itself".

The ominous creep of the hands of the Doomsday Clock has invaded human awareness since 1947, when this grim chronometer was first launched by the *Bulletin of the Atomic Scientists* as a means of alerting us to just how close to ultimate disaster we are (Bulletin of the Atomic Scientists 2015). The idea for the clock originated with a group of Manhattan Project researchers, responsible for building the world's first nuclear weapons, who called themselves the Chicago Atomic Scientists. After the destruction of Hiroshima and Nagasaki they began to circulate a small newsletter about nuclear issues which grew into the *Bulletin*. Co-founder of the publication Hyman Goldsmith invited artist Martyl Langsdorf, the wife of one of the nuclear scientists, to come up with a cover design for the June 1947 issue: she chose a clock to depict the imminence of nuclear peril to humanity. "The Bulletin's clock is not a gauge to register the ups and downs of the international power struggle; it is intended to reflect basic changes in the level of continuous danger in which mankind lives in the nuclear age...," another *Bulletin* founder, Eugene Rabinowitch, told the press. Or, as the media duly and more succinctly noted, "If the clock ever strikes 12, it's all over" (The Doomsday Clock 1984).

Originally the clock—which still hangs on a wall at the University of Chicago—depicted the nuclear threat alone; recently it has included global warming. Over the decades its minute hand has advanced and receded as the delicate balance between human wisdom and unwisdom ebbed and flowed. It started showing 7 min to midnight 1947, leapt to 3 min when the Soviets tested their first warhead in 1949, advanced to 2 min when the US built the world's first hydrogen bomb in 1952, receded to 12 min with the signing of the first nuclear test ban treaty in 1963, sprang to 3 min again during the iciest period of the Cold War in 1984, unwound to 17 min with the break-up of the USSR in 1991, and advanced back to 5 min in 2007 with the dawning of the 'second nuclear age'. In early 2015 its hands were re-set to 3 min to midnight (Bulletin of the Atomic Scientists 2016). "It is now 3 min to midnight," Kennette Benedict, the Bulletin's publisher told a news conference in Washington, DC. "The probability of global catastrophe is very high. This is about the end of civilization as we know it. In 2015, unchecked climate change, global nuclear weapons modernisations and outsized nuclear weapons arsenals pose extraordinary and undeniable threats to the continued existence of humanity. World leaders have failed to act with the speed or on the scale required to protect citizens from potential catastrophe. These failures of political leadership endanger every person on Earth" (Rice 2015).

Most credible scenarios for a self-inflicted end of civilisation, or indeed the human species, invoke the widespread use of nuclear and other weapons of mass destruction—but this in turn is most likely to occur as a consequence of other extreme stresses such as accelerated climate change, resulting famines, water and other resource crises, plague pandemics, refugee tsunamis and the failure of states and financial systems. Such a scenario illustrates the intimately connected nature of the various pressures building on humanity as a result of resource overuse (Chaps. 3 and 7), self-pollution (Chap. 6) and an out-of-control climate (Chap. 5). It underlines why each of these problems cannot be dealt with on its own, but only as components of the central existential challenge which faces humans in the C21st.

A wise humanity will recognise it isn't a case of overcoming a few, separate threats. All must be resolved together.

Tools for Armageddon

In the Biblical book of Revelation, Armageddon is the place of a gathering of kings in the end times—popular belief to the contrary, no mention whatever is made of battles. The Biblical version is thought by experts to have become entangled in folklore with the historical state of Megiddo (Har Megiddo in Hebrew, which would sound very like "arma", or weapons, to a Latin-speaker), a seven-thousand-year-old fortified town which was several times conquered by invaders from Syria, Egypt and elsewhere. The folk symbolism of 'Armageddon' as the last battle before the end of humanity thus served much the same purpose of forewarning people in the previous two millennia era about the existential dangers of warfare, as the Doomsday Clock has in more recent times. Both are grounded in the wisdom that humans have the innate capacity to obliterate themselves. This book presents an alternative (Chap. 10), a pathway for avoiding self-destruction.

Eight countries have the technical capability to unleash nuclear mayhem with significant stocks of operational weapons, according to the Arms Control Association. In 2015, of the five "official" nuclear armed states:

- China had about 260 total warheads.
- France had around 300 operational warheads.
- Russia had about 1512 strategic warheads deployed on 498 missiles and bombers and was thought to hold another 1000 strategic warheads and 2000 tactical nuclear warheads. Several thousand more awaited dismantlement.

- The United Kingdom had 160 deployed strategic warheads and a total stockpile of 225.
- The United States had 7700 nuclear warheads, including tactical, strategic, and stored weapons. These included 4500 active warheads and 3200 'retired' weapons (Arms Control Association 2015).

Three nuclear states—India, Israel, and Pakistan—never signed the Nuclear Nonproliferation Treaty (NPT), but are known to possess a nuclear arsenal. Of these ACA estimates:

- India had 120 nuclear warheads.
- Israel had 80 nuclear warheads (Kristensen and Norris 2014).
- Pakistan had 120 nuclear warheads.

Of the nuclear-capable states, the ACA says Belarus, Kazakhstan, and Ukraine held nuclear stocks following the collapse of the Soviet Union, but returned them to Russia and signed the treaty as non-nuclear-weapon states. South Africa secretly developed and then dismantled a small number of nuclear warheads and also joined the NPT in 1991. Iraq had an active nuclear weapons program before the 1991 Gulf War, but was forced to dismantle it under the eyes of United Nations inspectors. Libya voluntarily renounced its secret nuclear weapons efforts and Argentina, Brazil, South Korea and Taiwan also shelved their programs.

That leaves three putative nuclear powers: Iran, North Korea and Syria. Iran launched its nuclear energy program with help from the US in 1957. After the 1979 Iranian revolution the new regime continued to pursue nuclear energy for peaceful uses, and opposed the building of weapons on theological grounds. However, that view may have shifted—and by 2012 US intelligence sources had concluded "Iran has the scientific, technical and industrial capacity to eventually produce nuclear weapons" (Farnsworth 2014). It was suspected that Iran was running "ostensibly a peaceful nuclear energy program while reserving the option to make a political decision to build and deploy nuclear weapons."

North Korea tested its first nuclear device in 2006, announced it had a nuclear weapon in 2009 and by 2015 was estimated by ACA to have separated enough plutonium to build 6–10 warheads. Further underground testing was carried out in 2009 with experts variously assessing the yield at between 6 and 40 kilotonnes (from one third to nearly three times the power of the Hiroshima bomb). In 2012 North Korea announced it would suspend its program, but promptly aroused fresh global suspicion with the testing of a missile delivery system, leaving arms control talks stalled. Chinese estimates put its nuclear stocks at 20 devices in 2015 (Blair 2015) and the US-Korea Institute estimated they would hold 50–100 weapons by 2020 (Brunnstrom 2015). In the cases of both North Korea and Iran, the nuclear option serves a dual purpose—the usual threat and deterrence against potential foes and as a diplomatic tool to gain leverage in dealings with countries far more powerful than themselves, such as the US and China. As a number of observers have pointed out, one of the gravest risks in Iran acquiring nuclear weapons capability is that it may steamroller other Middle Eastern states into acquiring them, leading to an escalation of the fourteen century-old Sunni-Shia conflict into the nuclear realm. The potential savagery of such faith-based confrontations is graphically attested in Europe's 30 Years War (1618–1648) between Catholics and Protestants, and the Chinese Taiping Rebellion (1850–1864) between Christians and the Qing government which between them claimed 28–38 million lives. In religious conflicts, the assumption that the fear of 'mutually assured destruction'—which has prevented nuclear wars hitherto—will prevent holocaust, or that reason will prevail between competing forces, no longer holds good.

In September 2007, the Israeli air force bombed a target in Syria, believed to be the construction site of a nuclear research reactor similar to North Korea's Yongbyon reactor, the result of a 10-year collaboration between the two countries. Though a signatory to the nonproliferation treaty Syria subsequently failed to co-operate with the International Atomic Energy Agency on scrutiny, although its efforts to acquire other research reactors came to nothing and civil war claimed its attention (Wikipedia 2015). But, with Syria, Iraq, Iran and Pakistan combined, this leaves a lot of nuclear knowhow and technology in a region increasingly susceptible to terrorists, insurgents and religious fanatics.

Progress has been made in reducing the world's stocks of nuclear weapons and materials and in at least slowing the arms race through measures such as the Treaty on the Non-Proliferation of Nuclear Weapons, nuclear test bans and strategic arms limitation agreements. In an area shrouded in secrecy, the best estimates indicated that around 16,000 nuclear weapons or decommissioned warheads were still extant in 2015 (Arms Control Association 2015), down from around 60,000 at the height of the Cold War—but nevertheless still sufficient to exterminate humanity hundreds of times over. Opinions are sharply divided as to whether this reduction represents real progress in terms human safety, or whether the main players have been procrastinating and are insincere, or merely very cautious, when it comes to complete disarmament. Regardless, while there is general world agreement to eliminate chemical and biological weapons, for nuclear arms no such agreement exists.

According to Sharon Squassoni, director of the Proliferation Prevention Program at the Center for Strategic and International Studies, disarmament had by early 2015 "virtually ground to a halt".

The US was planning to spend a third of a trillion dollars on new nuclear weapons by 2025 while Russia was also upgrading its arsenal. The UK was continuing to develop its strategic nuclear submarine program, France was also building a next generation air-to-ground nuclear missiles and China was developing a new ballistic missile submarine. India was planning to expand its nuclear submarine fleet while Pakistan has started a third plutonium reactor and was developing a new short-range nuclear missile. Israel was modernising its nuclear strike force, and North Korea was pressing ahead with its weapons program. Any sense of optimism for disarmament that had prevailed after the end of Cold War had "essentially evaporated," Squassoni said (Santini 2015).

'Winter Is Coming…'

The most publicised horrors of nuclear war, over the past half-century, were blast damage, fireball burns and radiation sickness, as they were in Hiroshima and Nagasaki, leading to a perception that those well away from target areas might be spared. Scientists however demur, arguing that the biggest killer of all is likely to be a 'nuclear winter', triggered by the immense quantities of dust and smoke from burning cities and forests lofted into the upper atmosphere, and the simultaneous stripping of the Earth's protective ozone layer: "In the aftermath… vast areas of the earth could be subjected to prolonged darkness, abnormally low temperatures, violent windstorms, toxic smog and persistent radioactive fallout." This would be compounded by the collapse of farming and food production, transport, energy grids, healthcare, sanitation and central government. Even in regions remote from the actual blasts people would starve, die from freezing temperatures as much as 30 °C below normal, from radiation sickness and a pandemic of skin cancers, pollution and loss of immunity to ordinary diseases. The nuclear winter is in effect the antithesis of global warming, a shock cooling of the entire planet, but one lasting several years only. However, "A number of biologists contend the extinction of many species… - including the human species—is a real possibility," they say (Turco et al. 2012).

In the 1980s a group of courageous scientists[1] alerted the leaders of both the US and Russia to the dangers of a nuclear winter. In an atomic war, they warned, there will be *no* winners. Then-Soviet president Mikhail Gorbachev took their counsel to heart: "Models made by Russian and American scientists showed that a nuclear war would result in a nuclear winter that would be extremely destructive to all life on Earth; the knowledge of that was a great stimulus to us, to people of honor and morality, to act in that situation," he subsequently related (Hertsgard 2000). US President Ronald Reagan concurred: "A nuclear war cannot be won and must never be fought," he said in his State of the Union Address in 1984 (Reagan 1984). Marking this watershed moment in history Al Gore recounted in his Nobel Prize oration in 2007 "More than two decades ago, scientists calculated that nuclear war could throw so much debris and smoke into the air that it would block life-giving sunlight from our atmosphere, causing a 'nuclear winter.' Their eloquent warnings here in Oslo helped galvanize the world's resolve to halt the nuclear arms race."

How large a nuclear release is required to precipitate a nuclear winter is still subject to technical debate, but with the greatly improved models developed for climate science, recent estimates suggest as few as 50 Hiroshima-sized bombs (15 kilotonnes each) would do it—or the use of only one weapon in every 200 from the global nuclear arsenal (Robock 2009). This puts a very different complexion on the contemporary risks facing humanity.

First, it suggests that even a limited conflict among lesser actors in the arms race, for example between Pakistan and India, India and China or Israel and Iran, and involving mainly the use of "battlefield" nukes could still imperil the entire world. In *Lights Out: how it all ends*, nuclear experts Alan Robock and Brian Toon examined the effects of a regional war (Robock and Toon 2012). To begin with, they argue, a 'limited nuclear war' is highly unlikely as, with the release of a handful of battlefield nukes, things will very quickly spiral out of control as communications fail and panic spreads, mushrooming into a more general conflict involving dozens of weapons spread over a much wider region. Firestorms in the megacities would throw up a shocking amount of smoke, ash and dust—around 70 billion tonnes is the estimate for an India/Pakistan clash. Running this through climate models they found it would block out sunlight, chilling the planet by an average 1.25° for up to 10 years—enough to cause crop-killing frosts, even in midsummer. This would sharply reduce and in some regions eliminate farm production for several years. Normal world

---

[1] Among them Richard Turco, Carl Sagan and Stephen Schneider from the US and Vladimir Sergin and Vladimir Aleksandrov from the USSR (Turco et al. 1983; Aleksandrov and Stenchikov 1983).

grain stocks are sufficient to feed humanity for only about 2–3 months, so one of the first round effects of the war would be worldwide panic and financial collapse as food supplies give out and grain prices soar astronomically. A billion people living on the margins of hunger would probably perish within weeks, and billions more over the ensuing months.

In the early twenty-first century at least eight nations, on this calculus, have the tools to terminate civilisation, and possibly the human species, on their own, while at least two more aspire to the power to do so. Meanwhile the shadow of possible nuclear and chemical terrorism, and their consequences, is lengthening.

Nuclear Terror

The emergence of a gaggle of non-state players and terrorist groups—many of them impelled by belief systems that disavow civilised human values or are frankly delusional—has added a dangerous and volatile new ingredient into the nuclear cauldron. Quite simply, it means that the doctrine of 'mutually assured destruction' (MAD), which held at bay nuclear conflict between America and Russia for decades, no longer represents a reliable deterrent to the use of nukes, as there are now potential players for whom personal annihilation holds no significance.

The credibility of this new threat also derives from the existence of a very large global stockpile of weapons-grade material and tens of thousands of old Soviet and American warheads not yet destroyed which could potentially enter the black market. "The global stockpile of highly enriched uranium (HEU) as of the end of 2012 is estimated to be about 1380 ± 125 tons. This is sufficient for more than 55,000 simple, first-generation implosion fission weapons. About 98 % of this material is held by the nuclear weapon states, mostly by Russia and the United States," says the International Panel on Fissile Materials, adding "The global stockpile of separated plutonium in 2012 was about 495 ± 10 t" (International Panel on Fissile Materials 2013). Although the stockpiles of both weapons and materials have been coming down in recent years, they remain several orders of magnitude in excess of a theoretical '50 nukes' safety margin for the planet. Indeed, it is doubtful safety will ever be achieved, short of universal total nuclear disarmament. Even then, stockpiles of fissionable material would remain—and would leak stolen material. Evidence in support of this view can be found in International Atomic Energy Agency's Incident and Trafficking Database, which documents 2500 cases of theft or loss of nuclear materials, plus another 1400 cases of unauthorised use. Between 1993 and 2013, there were sixteen confirmed cases of the theft of plutonium or highly-enriched uranium—and those are only the ones we know about. Nuclear materials of all types are typically stolen *once every ten days* (International Atomic Energy Agency 2014).

In his book *On Nuclear Terrorism*, Harvard University's Michael Levi dissected the route an aspiring atomic terrorist would have to take to source, acquire, build and deliver a bomb (Levi 2007). It turns out it is not so straightforward, because there are a great many links in the chain and many things than can go wrong at each point. Also there are numerous far cheaper, simpler and more effective actions available to the pragmatic terrorist (assuming terrorists are pragmatic, not just crazy). For one thing, nukes and their ingredients are very costly, difficult to source and hard to construct without risking a horrible accident. Any organisation seeking them would have to be extremely well-funded to acquire a bomb, or the materials, components, equipment and technical skills required to make one. It would risk discovery at multiple points along the chain. However, Levi argues, the big, insecure stockpiles still extant globally nevertheless offer "gateways to terrorism". He concluded that numerous opportunities to detect and foil such an attempt also exist all along the chain—but that authorities must focus on all of the links in order to minimise the chances of a small, rich group of nuclear fanatics launching their own version of 'Armageddon'. While it seems implausible that terrorist groups, even an alliance, could ever manage to build and release 50 or more nuclear weapons, the possibility of them detonating one or two, which then precipitate a wider nuclear conflict among State players reacting in panic and leading to universal catastrophe, cannot be discounted.

Also, as Columbia University professor Jeffery Sachs has pointed out, 'terrorism' isn't always terrorism in the eyes of so-called 'terrorists'. It is the extension of an existing conflict to other battlefields, including the cities of the major powers themselves. This, he argues, is an understandable response to decades of western military violence, scheming and secret warfare against the peoples of the Middle East, often with oil as a motive (Sachs 2015a). If Sachs is correct and the 'terrorists' are mainly responding to western intervention by widening the theatre of conflict to include the intervening states themselves, then it follows that the risk of nuclear terrorism can also be de-escalated by the west ceasing its military involvement and political manipulations in the Middle East.

However "A… misimpression is that the problem has been solved by the end of the nuclear arms race," caution Robock and Toon: "The only way to eliminate the possibility of climatic catastrophe is to eliminate the (nuclear) weapons" (Robock and Toon 2012). In 2010, Jacob Kellenberger, president of the International Committee of the Red Cross, formally proposed "negotiations aimed at prohibiting and completely eliminating such weapons through a legally binding international treaty"—a universal ban on all nukes (Kellenberger 2010).

So far, nobody has taken him up on it. Indeed, the nuclear arms race is far from over. While countries may not be adding megatonnage to their nuclear arsenals, as they did in the 1960s and 1970s, they are piling up new technologies which are expected to increase the level of threat. Writing in the *Bulletin of the Atomic Scientists*, peace researcher Mark Gubrud, of the University of North Carolina, says "Hypersonic missiles are just one aspect of a renewed strategic arms race among the world's major nuclear-armed powers. Sources of this resurgent danger include smoldering geopolitical rivalries, shifts in economic power, and new weapons made possible by emerging technology. The world has failed so far to put the nuclear genie back in the bottle, and new genies are now getting loose: space weapons, cyber warfare, drones, and autonomous weapons. Weapons based on synthetic biology and nanotechnology loom on the horizon" (Gubrud 2015).

Chemical and Biowarfare

Chemical and biological weapons, while more limited in terms of their global impact than nukes, nevertheless pose a continued existential threat to humanity, despite recent progress in eliminating the world's stocks. The main weapons are nerve poisons (such as sarin, tabun and VX), blister agents (such as phosgene and nitrogen mustard), blood agents (like hydrogen cyanide) and lung agents (like chloropicrin and phosgene). Originally outlawed under the Geneva Convention in 1925, this didn't stop at least 20 countries from testing, making or stockpiling them.

Biological weapons consist of: lethal plague disease organisms like those which cause anthrax, botulism, cholera, encephalitis and haemorrhagic fever; natural deadly poisons like ricin or saxitoxin; various agents intended to disable, rather than kill, an opponent and their population; various herbicides, pest insects and fungal diseases intended to destroy or undermine agriculture and the food supply. Around seven countries have laboratories for either making or countering biowarfare agents, and eight more countries are suspect.

In 1992, under the Chemical Weapons Convention (CWC), it was agreed by participating nations to destroy the world's stocks within 10 years and 190 countries signed up. Initial declared stockpiles amounted to nearly 70,000 metric tonnes of toxic agents in 8.6 million munitions and containers, with the biggest arsenals being held by Russia (40,000 t) and the United States (30,000 t). The dismantling and destruction proceeded, with around 80 % of the total arsenal being verifiably destroyed although both the 2007 and 2012 deadlines for complete destruction were not met, with the main players seeking extensions on various pretexts (Arms Control Association 2014a). Israel and North Korea refused to declare their stocks.

In 2014 the Arms Control Association estimated that around 12,000 tonnes of nerve agents and four million bombs and shells still existed worldwide, sufficient to kill every human being two or three times over. These remaining stocks were concentrated in Russia (9000 kt) and the US (2800 kt) and were scheduled for complete destruction within 2–3 years (Arms Control Association 2014b). Two other countries, North Korea and Syria, are thought to still have them, and there are 16 suspects (Kerr 2008).

So, like nuclear arms, chemical and biological weapons have ceased to proliferate and have diminished in volume, number and killing-power over recent years—but they remain far from being totally abolished, destruction has proceeded much more slowly than hoped, and the shadow they cast on the human future has not been dispelled.

"The danger posed by biological weapons (BW) and chemical weapons (CW) still lingers two decades after the cold war's end. Despite the reduction of threats as an increasing number of states fulfil their commitments under international conventions, a small number of states still maintain declared and undeclared stockpiles and even active BW and CW programs," warns the Arms Control Association. "A bio-technology revolution is making bio-technology more readily available and presents a potential future proliferation risk. Dual-use chemical processes also present a series of ongoing challenges" (Arms Control Association 2014a, b).

Since an anthrax attack launched by an individual in the United States in 2001, authorities have regarded the risks of bio- or chemo-terrorism as higher than nuclear terrorism for the simple reason that biological agents and precursor chemicals are far cheaper and easier to obtain than nuclear materials and technology, and their manufacture is also technically simpler. Furthermore, if the main object is to generate fear and panic, they are just as effective. As a consequence, most developed countries have developed measures for detecting possible attacks while at world level the Global Counterterrorism Forum, comprising 30 countries, was launched in 2011 to improve international co-operation in surveillance and ways for dealing with this novel form of twenty-first century terrorism (Global Counterterrorism Forum 2016).

Increased Vulnerability

One reason why weapons of mass destruction are more to be feared in the twenty-first century than in the twentieth is that humanity is much more vulnerable than in the past.

By mid-century three quarters of the world's population will live in cities, which are invitingly concentrated targets for nuclear, chemical or biowarfare weapons. Modern cities also depend on highly fragile energy and transport chains to deliver power, food and water to dense populations every minute of the day: these would almost certainly fail immediately following an attack. This would in turn collapse healthcare, communication and emergency services and probably most forms of government. In the case of bio-weapons involving infectious disease, their impact will be magnified by modern travel, transport, education and child-minding systems, urban living conditions and work practices and by floods of refugees dispersing in all directions.

Conflict Drivers

We humans share a universal mythology, fostered by politicians, militaries and by—as historian and ex-warrior Paul Fussell once characterised them—"the sentimental, the loony patriotic, the ignorant and the bloodthirsty" (Fussell 1989), that war is noble and justifiable. However, this is not a view commonly shared by most citizens, especially by women and the educated, both male and female; with the gradual fading of the old imperialist tendencies, fewer and fewer swallow it. In democracies especially, the majority tend to oppose the military adventures of their governments. More than 200 million people have perished in the major bloodlettings of the last two centuries, by far the greater proportion of whom were non-combatants.[2] However, these wars are as a pale shadow to the loom of twenty-first century conflicts.

There is a brutal arithmetic to the politics of modern warfare between supposedly-rational participants—that the losses in blood and treasure should not be seen by the public at large to exceed the gains in land, wealth, security and national prestige. Where they do, the belligerent regime often finds itself out of favour with its own people. "Thus, for war to occur with rational actors, at least one of the sides involved has to expect that the gains from the conflict will outweigh the costs incurred," say economists Matthew Jackson

---

[2] For one estimate of fatal casualties in various conflicts see http://en.wikipedia.org/wiki/List_of_wars_by_death_toll.

and Massimo Morelli (Jackson and Morelli 2011). They identify the main drivers or motivators for contemporary warfare as: religion, revenge, ethnic cleansing and bargaining failure over resources.

While the media tendency to dramatise war invariably throws the spotlight on the ideological, religious, racial or political factors propelling the combatants, in reality disputes over resources have underlain or exacerbated most conflicts historically. Were we able to interview the combatants in that stone-age Gwion Gwion rock painting, it is likely they would tell us it was an argument over hunting rights or a water hole, that led to that historic discharge of weapons.

In the case of World War II, more mythologised that any conflict (Weber 2008), resources played a central role in precipitating war. As early as the 1920s, Hitler telegraphed his intention of taking large areas of Eastern Europe as '*lebensraum*' (living space for German farmers) in response to a feeling common among Germans at the time that there was a national overpopulation crisis. Subsequent histories concluded that for the Hitler and the Nazis, *lebensraum* was in fact their most important foreign policy goal (Messerschmidt 1990). German military strategy was also significantly dictated by the need to acquire oil and coalfields as well as farms in Russia, Romania and elsewhere. Japan—as an industrial and military economy—was also critically short of oil, depending for most of its needs on imports from the US. Acquiring its own oil supplies formed a central plank in its motivation for war and military planning, and led to its invasion of Indochina. When America countered with a total trade and oil embargo on Japan in July 1941, war between the two became inevitable, as the Roosevelt Administration duly recognised at the time (Children in History 2012). The pattern of Japanese conquest of southeast Asia and the Pacific islands was strategically driven by its need to acquire and defend oil, rubber, food and other resources from Indonesia, Malaya, southern China and the Philippines.

Up to half of all inter-state wars since 1973 have been linked to oil, says Jeff Colgan of the Harvard Kennedy School. "Although the threat of 'resource wars' over possession of oil reserves is often exaggerated, the sum total of the political effects generated by the oil industry makes oil a leading cause of war. Between one-quarter and one-half of interstate wars since 1973 have been connected to one or more oil-related causal mechanisms. No other commodity has had such an impact on international security," he says. Colgan identifies eight different ways in which oil helps precipitate, stoke or underpin conflict and warns that the number of security concerns is multiplying as new oil exporters enter the global market (Colgan 2013). It follows that ceasing to use oil will remove a major driver of conflict.

In 1999 the Oslo Peace Research Institute issued a ground-breaking paper by Indra de Soysa and Nils Gleditsch which drew attention to the fact that, in the first decade of the post-Cold War era, most conflicts began with development failure and contests between the different players over those fundamental resources for life: food, land and water. "The new internal wars, extremely bloody in terms of civilian casualties, reflect subsistence crises and are largely apolitical," they said (De Soya and Gleditsch 1999). This represented a challenge to the long-held academic view that scarcity is a product of war—rather than war a being product of scarcity. In fact, humans have always contested key resources *vi et armis*—and politics, religion, patriotism and ethnicity are just the way we tend to marshal ourselves into opposing groups around them. Peter Gleick's work on water conflicts lends substance to the warnings of two UN chiefs, Boutros Boutros-Ghali and Ban Ki-Moon, of the increased danger of wars breaking out over this indispensable resource as scarcity takes hold. 'Food wars' (including so-called 'fish wars') have erupted on numerous occasions in Africa—where the Rwandan genocide and drawn-out bloody conflicts in Darfur and the Horn of Africa are particular examples—but also in Central America and Asia (Messer et al. 1998). These fights are almost always over the fundamentals of human survival and tend to originate as civil conflicts, which then spiral out of control to embroil neighbour states and even the superpowers.

From the depth of his experience as both a farmer and an international statesman, former US president Jimmy Carter observed that modern wars almost invariably begin in poor countries where resources and people are stressed—seldom in rich ones or in democracies. Writing in the International Herald Tribune, he said "The message is clear. There can be no peace until people have enough to eat. Hungry people are not peaceful people" (Carter 1999a).

In the emerging era of resource instability, described in Chap. 3, the risk of war is liable to increase in proportion to the scarcity of essential resources, be they water, farm land, food itself, oil, gas or strategic minerals. The possibility that some of these conflicts will involve the discharge of chemical, biological or nuclear weapons cannot be discounted. For example, in their *Age of Consequences* report, Kurt Campbell and colleagues at the US Center for Strategic and International Studies (CSIS) foreshadowed that with the famines and global disruption arising out of severe climate change (2.6 °C, in their scenario) "It is clear that even nuclear war cannot be excluded as a political consequence. Moreover, so-called "limited nuclear war" in any part of the world can escalate to a full-scale nuclear exchange among the big nuclear powers." With catastrophic change of 5° or more, "The probability of conflict between two destabilized nuclear powers would seem high." Furthermore "Armed conflict between nations over resources and even territory, such as the Nile and its tributaries, is likely, and nuclear war is possible" (Campbell et al. 2007).

The report also raises a disturbing new dimension of nuclear proliferation—the fact that nuclear power generation is increasingly being touted worldwide as a 'climate friendly' replacement for carbon-emitting coal-fired power stations. In many countries it is also justified for the desalination of sea water or to drive economic growth. A standard 'peaceful' nuclear reactor, over its lifetime generates sufficient enriched material to make 1200 Hiroshima bombs (Nuclear Information and Resource Service 1996). If the number of nuclear power plants worldwide were to double to one thousand or more in a bid to slow global warming, they would together produce enough fissionable material to construct over a million weapons. The claim that every kilo of this waste will remain in safe, responsible custody *forever* and that none will leak out to terrorists, religious fanatics or ambitious and ruthless minor powers is refuted by the frequent thefts and losses documented by the IAEA. "Nuclear material is stolen or lost two to four dozen times a year every year. Sometimes small amounts, sometimes large. It happens an awful lot in Russia and other former Soviet states; it happens in poorer, nuclear-capable countries such as Mexico, India and South Africa; and you'd better believe it happens in rich countries, as well, particularly in France," the Washington Post reported (Fisher 2013).

In other words, adopting the 'nuclear solution' to climate change may only hasten the end of civilisation—by other means. It is not, therefore, a wise solution.

'Armageddon' Scenarios

Contemporary scenarios for the endgame in human history include:

- A miscalculation by either party, or both, in the growing tensions between a US striving to reassert its superpower status and a resurgently nationalistic Russia operating to reclaim its former sphere of influence.
- An unpredictable escalation in the political tensions, territorial and resources disputes in East Asia, involving several countries and drawing in the US, China and possibly Russia.
- The US, Russia or China invades a smaller nation or occupies new territory, provoking military retaliation by one or both of the others, escalating into nuclear conflict.
- A misunderstanding or terrorist attack triggers nuclear conflict between India and Pakistan.

- A 9/11-type nuclear strike against any of the nuclear powers (US, Russia, China, India, France, UK, Israel) by a terrorist or dissident group, leads to confusion, nuclear retaliation against perceived 'host' countries and escalation.
- Conflicts over resource scarcity and refugee tsunamis precipitated by climate change, escalating into local nuclear wars.
- The rise of nuclear tensions and capabilities in Israel and Iran, leading to a miscalculation or a deliberate first strike by one to neutralise the other, then escalation involving other actors.
- The risk of a Sunni-Shia nuclear conflict arising out of a Middle East arms race precipitated among other Arab states by Iran acquiring nuclear weapons.
- Random use of nuclear or chemical devices by terrorist groups prepared to risk all, or delusional about the consequences, leading to global panic, confusion and retaliation.
- Use of mini-nukes and dirty nukes in internecine contests between competing terrorist or religious fundamentalist factions, spiralling out of control. A particular scenario is that rising confrontation between Moslem Sunni and Shia actors, and the nations that back them, escalates into a nuclear conflict with global consequences.
- Use of WMDs by nuclear-capable states, or even non-state players including powerful criminal or religious groups, in escalating disputes arising over scarcities of key resources including food, land, water, minerals or energy (Moore 2007).
- Use of WMDs or nukes to control refugee tsunamis flooding out of climate- or war-ravaged lands.
- A malfunction in a robotic nuclear device, leading to an unintended attack on a nuclear state, which retaliates.

From this list it can be seen that the principal risks of nuclear conflict fall into two categories—the lesser being disputes over physical resources such as water, oil and land, and the greater being political, religious, nationalistic or ethnic differences.

In short humanity is somewhat more likely to destroy itself over something imaginary—such as a belief, a border or a political theory—than it is over something real. However, each constitutes a realistic *casus belli*. This accentuates the importance of species wisdom in preventing such circumstances from ever arising.

Unheeded Warnings

Despite significant progress in reducing stocks of both nuclear and chemical weapons since the 1980s, neither have been abolished and in the case of nuclear arms especially, there is no global agreement to do so.

The world scientific and medical communities express mounting frustration and alarm that their warnings, once taken very seriously by world leaders, now seem to fall on deaf ears. "To our great peril, the scientific community has had little success in recent years influencing policy on global security," Lawrence Krauss, a physicist from Arizona State University and co-chair of the Bulletin of Atomic Scientists, lamented in the New York Times. "Scientists' voices are crucial in the debates over the global challenges of climate change, nuclear proliferation and the potential creation of new and deadly pathogens. But unlike in the past, their voices aren't being heard," he said, going on to explain that scientific advice on nuclear disarmament was 'routinely ignored'. "Until science and data become central to informing our public policies, our civilization will be hamstrung in confronting the gravest threats to its survival," he concluded (Krauss 2013).

"Back in the days of the Cold War, people were terrified that there would be a nuclear war. Guess what? The danger is still here and it is growing. We are living on borrowed time," say International Physicians for the Prevention of Nuclear War (IPPNW) (International Physicians for the Prevention of Nuclear War 2016).

As to the chances of a catastrophic nuclear war (one with a death toll exceeding 80 million), a group of experts polled by the Project for Study of the 21st Century in 2015, concluded there was a 6.8 % likelihood within the next quarter century. The chances of an act of nuclear terrorism were rated around 17 % (PSI 2015).

Nobel Peace laureate Dr Ira Helfand, who campaigns for a complete worldwide ban on nuclear arms, says "The problem with nuclear weapons is that people have tended to forget they are there. If just 100 were used there would be enough disruption to lead to global famine and put two billion people at the risk of starvation."

"If a major city, such as New York, were hit, 15 million people would die within half an hour," he said in an impassioned address to the 2014 Nobel Peace Laureates Summit. Following a general nuclear conflict "For three years… there would not be a single day free of frost. Ecosystems would collapse. Food production would stop. And the vast majority of the human race would starve to death. It is possible we would become extinct as a species.

"This is not some nightmare scenario I have cooked up. This is the danger we face every day that these weapons continue to exist. Nuclear weapons are not a force of Nature. They are not an Act of God. They are something that human beings built—and we can take them apart. We know how to do that. All that's missing is the political will. And for that we turn to you…" (Helfand 2015).

What We Must Do

1. Outlaw and destroy all nuclear weapons and stocks of nuclear materials.
    Pathway: it is down to the citizens of the eight countries with the ability to destroy civilisation, especially, to put pressure on their governments to walk away from the risk of universal destruction. Without citizens taking prime responsibility, it is doubtful if national governments, politicians and militaries ever will. It is up to the rest of us to support them in this cause.

2. Convert from uranium-based nuclear energy to safer systems (e.g. thorium and renewables) less able to produce weapons.
    Pathway: the modern uranium reactor was designed to produce the materials for nuclear weapons, even though it has since been adapted to produce electricity. It still produces fissile waste, so its use is a continuing threat to the human future and it should be eliminated for the same reasons we need to eliminate fossil fuels or pandemic diseases. Only citizen demand and political pressure for clean energy, including clean nuclear energy, can achieve this.

3. Outlaw and destroy all chemical and biological weapons and stocks.
    Pathway: again, informed citizen action is the route to safety, coupled with greater international efforts to build trust, mutual dependency and reduce the risk of conflict between nuclear-armed states.

4. Develop stronger, more collaborative global surveillance of nations and groups who pose a potential risk of WMD terrorism.
    Pathway: stepped-up international co-operation in intelligence sharing and efforts to discourage groups who may wish to use this threat by redressing their grievances, as well as by police action where needed.

5. Further develop a global citizens' movement operating in all countries and societies to warn of the dangers of continued retention of WMD and exert political pressure for their abolition.
    Pathway: use social media to reinvigorate and mobilise the worldwide disarmament movement.

6. Make a stronger national and international investment in conflict resolution.
   Pathway: strengthen current global institutions for peacemaking and resolving disagreements.

7. Earmark a fixed percentage of the global military budget for addressing global challenges liable to lead to war. Especially, earmark 10% of global military spending to 'peace through food'—i.e. ensuring a universal food supply adequate to reduce the tensions that lead to conflict.
   Pathway: see Chap. 7.

What You Can Do

- Understand that a nuclear inferno is an omnipresent threat to you, your children and to all posterity. It exists 24/7. The fact it hasn't happened in the last 70 years doesn't make us safe in the next 70. The risk is greater than at any time since the end of the Cold War.
- Actively support responsible citizen campaigns to ban nuclear, chemical and biological weapons and demolish stockpiles in your country and globally.
- Vote for no politician who does not commit to complete nuclear, chemical and biowarfare disarmament and stocks destruction.
- Share your views on this issue with friends around the world via social media and the internet. Spread the message.
- Avoid belief systems which encourage opposition, discrimination or hatred for other groups.
- Teach your children that, in the twenty-first century, with a precarious balance between human numbers and stressed resources and ecosystems, conflict is the road to ruin. There are no 'winners'. Co-operation and mutual understanding, on the other hand, is the road to peace.
- Support the resolution of conflict by peaceful means. Oppose the spread of all weapons.

5

The Baker (*Homo pistor*)

I hear hurricanes a-blowin'.
I know the end is coming soon.
I fear rivers overflowing.
I hear the voice of rage and ruin.
— Creedence Clearwater Revival, Bad Moon Rising, 1969

Slicing through the azure-tinted waters of the summer Laptev Sea, north of Siberia, the research icebreaker *Oden* encountered huge patches of ocean fizzing gently, like soda water. "This was somewhat of a surprise," wrote expedition leader Örjan Gustafsson of Stockholm University. In his view the bubbles emanating from the continental slope beneath the vessel originate with melting deposits of frozen methane gas in the seabed. Normally trapped by pressure and cold, these are now escaping as the temperature of waters in the Arctic Ocean climbs (Papadoupolou 2014).

Cruising along a depth gradient of between 100 and 1000 m on the slope and employing acoustic and geochemical testing, the Swedish and Russian scientific team detected "vast methane plumes escaping from the seafloor at depths between 500 m and 150 m". In several places, the methane bubbles were seen to breach the ocean surface. Chemical analysis of seawater samples pointed to levels of dissolved methane from ten to fifty times above normal.

Taken on its own, it was a mere curiosity—a few bubbles, seemingly of little consequence. Far from the first, quiet footfalls of planetary disaster. But what the *Oden* team witnessed was no isolated event: it was one beat in a rising drumroll. Only week or two earlier, in the record-breaking heat of the 2014 Arctic summer,

© Springer International Publishing Switzerland 2017
J. Cribb, *Surviving the 21st Century*, DOI 10.1007/978-3-319-41270-2_5

a giant hole had mysteriously appeared in the Siberian tundra. The crater, measuring 80 m wide and deep, initially drew global media attention as a bizarre and rather entertaining geological oddity; a rash of bizarre theories, from aliens, to meteor impacts to sinkholes, were offered to explain it. One of the strange features was a ring of erupted earth around its rim, apparently spewed out, pointing to some internal pressure suddenly and explosively released. Russian scientists concluded the hole was excavated by a combination of melting permafrost and an eruption of the methane gas it had contained, previously frozen solid in the Siberian tundra for tens of thousands, possibly millions of years. What seemed another unique event, turned out not to be: similar holes were quickly discovered across Siberia, a region of the planet where average temperatures have already risen by 2 °C (compared with the whole Earth's +0.85°) in the last half century. Russian researcher Marina Leibman said "I would argue this is a new process, which was not observed previously. It can be seen as a reaction to changes in the temperature, which releases gas, possibly hidden in the form of relic hydrate, from the upper layers of permafrost" (Liesowska & Lambie 2014).

Methane is a gas with many times the climate-forcing power of carbon dioxide—meaning it has the ability to cook the planet far more rapidly than the burning of fossil fuels. Scientists estimate there may be as much as five trillion tonnes of it locked up in permafrost on land (Schuur et al. 2015) and in shallow marine deposits, known as hydrates or clathrates, on continental shelves around the world. This methane is the accumulated waste from hundreds of millions of years of decomposing plant matter, algae and plankton in the oceans and swamps—a vast global compost heap, as large as or larger than all the coal, oil and gas deposits ever found, and formed by similar biogeological processes. The methane gas, identical to the bubbles that rise to the surface of a stagnant pool when the mud is stirred, has remained immobile, frozen within the seabed or tundra sediments by intense cold and pressure. Now, as the Arctic Ocean and landmasses warm, these vast natural deposits are starting to melt and vent, compounding the planet-warming effects of the carbon in the atmosphere released by human activities.

Scientific reports of the escape of methane have multiplied. A Canadian study, running since the mid-1980s, tracks an 8% overall increase in atmospheric methane—but also reveals a series of sudden, sharp spikes in escaping gas, well outside the trend. Scientists have given these harbingers the foreboding title of 'dragon's teeth', the first fiery breaths of runaway global warming. Other reports suggest the terrestrial carbon, at least, is currently venting at a steady rate and in moderate amounts. However, the big one is seafloor methane.

As Canadian glaciologist Dr Tim Box succinctly informed his Twitter followers: "If even a small fraction of Arctic sea floor carbon is released to the atmosphere, we're f'd" (Box 2014).

Global Warnings

The mechanism that drives global warming has been understood by science since at least 1896, when the Swedish scientist Svente Arrhenius discovered it while attempting to take the surface temperature of the moon. Arrhenius was measuring the infra-red (heat) radiation reflected from the lunar surface and noticed that his measurements were much more precise when the moon was high in the sky, than when it was low on the horizon, where the faint infra-red signals had to pass through far more of the Earth's atmosphere to reach the observer. The work of earlier scientists had pointed to the ability of certain gases, such as carbon dioxide, to trap heat. From this Arrhenius deduced that the moon's infra-red heat signal was being absorbed by gases in the atmosphere, specifically carbon dioxide and water vapour. From this observation he went on to calculate that halving the volume of CO_2 in the Earth's atmosphere would cause the planet to release heat, resulting in an ice age—whereas doubling it would store sufficient heat to lift global temperatures by 5–6 °C. It was the birth of 'greenhouse theory', though at the time it was considered more a scientific curiosity than something that would rule the fate of civilisation. Arrhenius recognised that emissions of CO_2 from human industrial and domestic coal burning would also tend to raise the Earth's temperature—but with the relatively small volumes being burnt in the 1890s, considered this process would take hundreds of years.

The process was investigated, off and on, for the next half century though it wasn't until the 1960s that scientists began to pick up clues that various human activities—pollution especially—were changing the physical behaviour of the atmosphere. This discovery led to far closer scrutiny and to the birth of modern climate science. At that time, the prevailing concern was that humanity's emissions of smoke and soot particles might pitch the climate back into a cooler era, a theory supported by the observation that major volcanic eruptions which pushed gases and fine particles into the upper atmosphere tended to be followed by cool years as a result of the 'sunshade' effect. However, by the early 1970s, more and more temperature readings were showing that what was actually happening to the globe was warming—not cooling. The media of the day was more entranced by the idea of a new Ice Age, than by something that sounded almost like good news. Thus, with scientists saying one thing and the media apparently the opposite, the public and legislators were pardonably confused—a dilemma which has persisted into the twenty-first century.

By the mid-1970s, however, the early climate models were pointing unambiguously to the conclusion that a doubling in atmospheric CO_2 levels would raise global temperatures by at least 2 °C. The phenomenon was formally christened by Wallace Broecker, a researcher at Columbia University's Lamont-Doherty Earth Observatory, on August 8th, 1975, in a paper in *Science* magazine which posed the question "Climatic change: are we on the brink of a pronounced global warming?" (Broecker 1975). Meanwhile other researchers were compiling proofs that various 'feedbacks' in the Earth system—the extent of icecap cover, the rate at which the oceans absorb and release CO_2, cloud cover, polar melting etc—could have major influences, some driving the process faster, others counter-acting to slow it down. By 1979 the trend of evidence was sufficiently clear for the World Climate Congress to conclude that humanity's rising CO_2 emissions were heating the lower atmosphere. It was the birth of a worldwide scientific consensus that has only strengthened with the accumulation of vast amounts of physical evidence, the refining of climate prediction models and deeper insight into how the Earth system functions.

The Venus Syndrome

Hanging in the sky each clear sunset like some blazing omen in the ancient heavens, is the planet Venus: the Evening Star has been an object of awe and veneration to humans down the ages. Every person with a working pair of eyes has seen it. Yet almost everyone has failed to read its foreboding message.

For this beautiful pearl-like object is a blistering 462 °C on its surface, hot enough to melt lead[1]; its super-dense atmosphere exerts 92 times the pressure of our own and is a lethal blend of carbon dioxide and sulphuric acid. It is baking, naked rock, devoid of water and life. All this is the result of a convulsion which has turned Earth's sister planet into something unpleasantly reminiscent of religious visions of a fiery Hell.

Earth scientist James Hansen has been obsessed with Venus, and how it became like it is today, since he joined NASA's Goddard Institute for Space Studies as a young researcher. Of the three earthlike planets—Mars, Earth and Venus—only Earth enjoys the 'Goldilocks climate', he says, one that is just right for life to arise and flourish in. Mars is too cold—minus 50 °C on a good

---

[1] The melting point of lead is 327.5 °C, 621 °F

day—and Venus is way too hot. These dramatic contrasts, Hansen concluded, were driven chiefly by two factors—the planet's albedo, or the fraction of its heat it reflects back into space, and the blanket of heat-trapping gases in its atmosphere. The atmosphere of Mars is extremely thin with little insulating power, so most of the planet's heat is shed immediately, making it freezing cold. Earth traps some incoming heat from the sun (about 71%) and reflects the rest (29%), which maintains a comfortable 'greenhouse' in which life can thrive. "Venus has so much carbon dioxide that it has a greenhouse warming of several hundred degrees," Hansen wrote (Hansen 2009). He considers this came about as a result of a runaway warming process in which the planet's temperature rose to 100 °C, whereupon its oceans boiled off into space and the surface then became so hot that all the remaining carbon was jettisoned into the atmosphere, forming the present planetary blast-furnace. It was his reading of this grim celestial omen that drove Hansen to risk his own career to alert his fellow humans to what might happen to the Earth if we stoke up the atmosphere with too much carbon.

Systematically, he set about advising fellow scientists, the US Government, the fossil fuels industry and the public. In 1988 he told the US Congress that human carbon emissions had already measurably affected the world's climate. The same year an international atmospheric physics conference concluded the changes were "a major threat to international security... already having harmful consequences over many parts of the globe." It warned that, to avoid this, humanity should seek to reduce its total carbon emissions to a level 80% of that prevailing in 1988 (World Meteorological Organization 1988). Hansen's outspoken campaigning infuriated the George W. Bush White House and antagonised America's powerful fossil fuels lobby. Steps were taken to gag him (Hansen 2009). They didn't work: Hansen became if anything more outspoken. He took part in public protests and was several times arrested. He offended a few fellow scientists by crossing the unspoken line between dispassionate science and passionate advocacy. He criticised industry and the US government for doing too little, too late. He demanded that coal and oil company executives be put on trial for 'high crimes against humanity and nature'.

Becoming a grandfather who cared about his grandkids' future and incensed by a throwaway remark by TV host Larry King that "nobody cares about fifty years from now", Hansen launched into a personal account, *Storms of My Grandchildren* (Hansen 2009). Part-science and part political philippic the book lays out the scientific evidence for global warming—and the negligent reaction of the US Government. However, it concludes optimistically, trusting that people who care about their grandchildren will act on the facts about warming. But in it Hansen also warns that, if we burn all the oil, coal and tar sands available to us, "the Venus syndrome is a certainty".

Hansen's was hardly a lone voice: today you can stock a respectable library with books about global warming theory and its ramifications. In 1989 Bill McKibben released *The End of Nature*, generally hailed as the first major account for general readers about the consequences of man-made climate change. In 2005 Australian biologist Tim Flannery published *The Weather Makers,* a graphic plain-language narration of the consequences humans will reap from our interference in the climate. The following year, former US vice president Al Gore launched his celebrated documentary and book *An Inconvenient Truth* and British environmental journalist George Monbiot published *Heat: how to stop the planet from burning.* In 2007 Mark Lynas put out *Six Degrees: Our Future on a Hotter Planet.* Then in 2008 British biologist Peter Ward released *Under a Green Sky,* which describes global warming episodes of the past and the mass extinctions they caused, Canadian Gwynne Dyer foreshadowed the geopolitical aftermath in *Climate Wars* and New York Times writer Tom Friedman appealed to America to lead the way out, in *Hot, Flat and Crowded.* In 2009 the celebrated English scientist James Lovelock, originator of the *Gaia* theory, wrote *The Vanishing Face of Gaia: a Final Warning* in which he concluded, darkly, "No human act can reduce our numbers fast enough even to slow climate change… We do not seem to have the slightest understanding of our plight…. Our wish to continue business as usual will probably prevent us from saving ourselves" (Lovelock 2009). In 2010, Australian academic Clive Hamilton published *Requiem for a Species* in which he dissected the extraordinary reluctance of politicians, industry and the public to take action over a clear and present danger to their future, while American historian Naomi Oreskes exposed the deceitful role of the fossil fuels lobby in undermining public confidence in climate science, in *Merchants of Doubt.* With these and scores of other books, TV programs, movies, TED talks, hundreds of official reports, more than 12,000 scientific papers and tens of thousands of media stories, humanity can scarcely claim we haven't had fair warning about the dangers of burning fossil fuels and liberating carbon.

The Facts

In 1989 the United Nations Environment Program and World Meteorological Organisation decided that the situation was sufficiently grave to justify forming a unique international task force to take forward world understanding of climate issues using peer-reviewed scientific findings—the Intergovernmental Panel on Climate Change (IPCC).

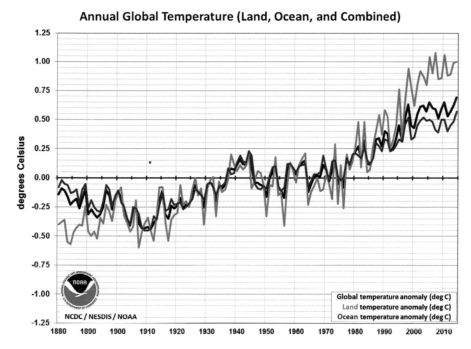

Fig. 5.1   Scientific measurements from around the world show an upward trend in global temperatures since the industrial age began, consistently setting new records early in the twenty-first century. *Source:* NOAA/(NASA 2015)

For the next quarter century thousands of researchers in hundreds of universities and research institutions around the world worked together to document and interpret the changes in the Earth's climate which were piling up in billions of measurements. The data—whether measured on land, in the air, in the oceans, or at the interface in the form of sea-level rise, told the same story: there has been a steady rise in the Earth's temperature. 2014 was officially proclaimed the warmest year on record—at 0.69 °C hotter than the average for the whole twentieth century—only to be eclipsed by 2015, according to the World Meteorological Organisation (WMO 2016). In February 2016, the world was shocked by reports that the surface of the Earth north of the equator was already 2 °C warmer than pre-industrial temperatures—this was the line that was never supposed to be crossed (Dyke 2016). Nine of the ten warmest years ever recorded occurred during the twenty-first century, and average temperatures rose worldwide for 38 consecutive years since 1977 (NOAA 2014). This information, and the scientific sources on which it is based, are freely available to any literate citizen of Earth with internet access and a care for their children's future (see Figs. 5.1 and 5.2).

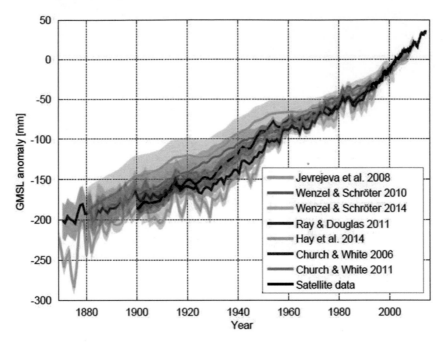

Fig. 5.2 The Earth's own 'thermometer': numerous scientific studies show global sea levels have risen steadily since the 1880s, due to a combination of melting polar icecaps and thermal expansion in the upper layer of sea water. While the studies report different findings from place to place, they agree about the trend. *Source:* IPCC (2014a)

In the light of the accumulating evidence, the IPCC has steadily ratcheted up its warnings about the perils of climate change since its first synthesis report appeared in 1990. In its fifth report (November 2014) it concluded:

- "Human influence on the climate system is clear, and recent anthropogenic emissions of greenhouse gases are the highest in history. Recent climate changes have had widespread impacts on human and natural systems.
- "Warming of the climate system is unequivocal, and since the 1950s, many of the observed changes are unprecedented over decades to millennia. The atmosphere and ocean have warmed, the amounts of snow and ice have diminished, and sea level has risen.
- Human carbon emissions are "extremely likely to have been the dominant cause of the observed warming since the mid-20th century"
- Some extreme changes in weather and climate "have been linked to human influences"

- Atmospheric concentrations of greenhouse gases are the highest they have been in 800,000 years.
- "Continued emission of greenhouse gases will… increas(e) the likelihood of severe, pervasive and irreversible impacts for people and ecosystems."
- More floods and heatwaves are likely.
- Climate changes will amplify existing risks and create new ones for humanity and the natural world
- There will be large risks to world food security.
- These effects will last for centuries, even if emissions cease (IPCC 2013).

What this means for our climate future is explained by scenarios known as RCPs (Representative Concentration Pathways). For example, Fig. 5.3 shows the difference between a scenario (RCP 2.6) in which we stop emitting carbon in the 2020s, emissions peak and then begin to decline—and one in which they continue to rise in line with growth in demand for fossil energy (RCP8.5). In the first case the planet will be around 2 °C hotter—and in the second around 5 °C hotter.

For humans, wise or unwise, the take-home message is that the temperate world in which our civilisation was born, developed agriculture, cities, industry and high technology, is gone. It will never return. We may still have a choice about what sort of planet we end up with, but even that window is closing rapidly.

Ours is a new and uniquely perilous age: just how perilous is not yet well or widely appreciated.

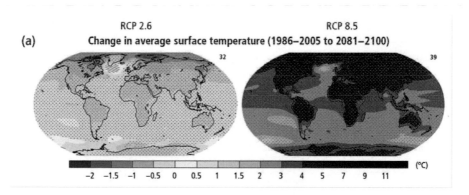

Fig. 5.3 Baked world: the impact of our continuing to burn fossil fuels ad lib (RCP8.5), compared with successful efforts to check carbon emissions (RCP2.6). *Source:* IPCC (2014b)

Keeping Our Cool

To have any chance of keeping our cool—to a global rise of 2 °C or less—humanity must abandon its addiction to fossil fuels *completely*. There is no longer any serious question of this. A crisis point is fast approaching, beyond which lies a path that could abolish civilisation.

That point will be the release by humans of a grand total of 2900 billion tonnes of CO_2 into the atmosphere, beyond which there will be no chance of retaining the 'warmer Earth' of the first scenario above, or of preventing the much hotter Earth of the second scenario. Since the industrial age began we have already released 1900 billion tonnes, so we're two thirds of the way to the crisis point. Since the late 2010s, with recovery from the Global Financial Crisis, the world has added an extra 50 billion tonnes of greenhouse gases (in CO_2 equivalents) to the Earth's atmosphere every year[2]. Assuming stable world economic growth, this means the 'point of no return' for a hot world will probably be reached during the 2030s.

If we are not to bake, then most of our known reserves of fossil energy must stay in the ground say Christophe McGlade and Paul Elkins of University College London: "A third of all oil reserves, half of all gas reserves and 80% of all coal reserves should remain unused from 2010 to 2050, in order to meet the target of 2°C." That includes no use of unconventional oil (from tar sands, coal or shale) or Arctic oil whatsoever, they added (McGlade & Elkins 2015).

Another way to express the danger point is in the concentration of greenhouse gases in the Earth's atmosphere. This began the industrial age at around 260 parts of CO_2 equivalent per million parts of atmosphere. In the 1950s it crossed 300 ppm and the planet was starting to warm perceptibly. In the early 1980s it attained 350 ppm and by 2015 it had reached 400 ppm and was rising by a steady 2 ppm each year. According to the IPCC, if we want to limit the planet's temperature to a 2 °C increase, the concentration of CO_2 must *never* be allowed to climb above 450 ppm. However, many scientists disagree, arguing that 350 ppm, or even 300, is the upper limit for a 'liveable Earth'. For example, James Hansen and colleagues declared "If humanity wishes to preserve a planet similar to that on which civilization developed and to which life on Earth is adapted, paleoclimate evidence and ongoing climate change suggest that CO_2 will need to be reduced… to at most 350 ppm… An initial 350 ppm CO_2 target may be achievable by phasing out coal use except where CO_2 is captured and adopting agricultural and forestry practices

---

[2] see for example, http://www.epa.gov/climatechange/science/indicators/ghg/global-ghg-emissions.html or http://www.ipcc.ch/publications_and_data/ar4/wg3/en/spmsspm-b.html

that sequester carbon. If the present overshoot of this target CO_2 is not brief, there is a possibility of seeding irreversible catastrophic effects" (Hansen et al. 2008).

If we don't succeed in cutting our carbon emissions, the World Bank has warned, we are heading for:

- "a world of climate and weather extremes causing devastation and human suffering"
- Extreme heat waves, especially inland, with temperatures from 4–10 °C hotter than at present. Average summer temperatures will rise by 6 °C or more.
- Sea levels rising by 0.5–1 m by 2100, exposing low-lying cities in South, East and SE Asia, Africa and Latin America to chronic flooding and forcing the abandonment of some islands.
- High temperatures that will damage yields of rice, wheat, maize and other important crops
- Agricultural food supplies worldwide will be disrupted, with famines in some regions and large fluctuations in both price and supply for all.
- Acute water scarcity in dry regions, affecting cities, food and fish production. Increased flooding in the tropics.
- At least 80 % of the world's population will be directly affected by one or more of these impacts (World Bank 2013).

"A 4 degree warmer world can, and must be, avoided—we need to hold warming below 2 degrees," declared World Bank Group President Jim Yong Kim. "Lack of action on climate change threatens to make the world our children inherit a completely different world than we are living in today. Climate change is one of the single biggest challenges facing development, and we need to assume the moral responsibility to take action on behalf of future generations, especially the poorest" (World Bank 2012).

Uncontrollable Warming

In 2009, Richard Zeebe of the University of Hawaii and colleagues made a profoundly disturbing discovery while checking out an event that took place some 55 million years ago—when the Earth took a sudden fever and its temperature shot up by 5–9 °C.

This event, known as the Palaeocene-Eocene Thermal Maximum or PETM, happened only about ten million years after the dinosaurs were smashed by an asteroid impact. This 'hyperthermal' period took place quite suddenly (in geological terms)—in less than 2000 years—and lasted for about 170,000 years before the planet again cooled. The heat spike was accompanied by a major wipe-out of ocean life in particular, though most small land mammals survived. Investigating the records of old marine sediments Zeebe was able to show there had been a sharp, 70%, leap in atmospheric CO_2 concentrations at the time. However, he concluded there was only sufficient carbon available to force the climate to warm by 1–3 °C and that some other mechanism must have been triggered by the initial warming, which then drove the Earth's temperature to fever pitch, up by another 4–6 °C (Zeebe et al. 2009). This process is the 'runaway global warming' which now menaces us.

The significance of PETM is that it appears that about the same volume of carbon was dumped by natural processes into the Earth's atmosphere and oceans as humans are currently dumping with the burning of fossil fuels and clearing of the world's forests—about 3 trillion tonnes in all—and it was this that triggered the hyperthermal surge in planetary heating.

As to the mechanism that could suddenly release a huge amount of extra carbon into the atmosphere and oceans and project global temperatures up by 6–9 °C, the most likely explanation is the one described at the start of this chapter—the rapid melting and escape of billions of tonnes of frozen methane, CH_4, currently locked in tundra and seabed sediments. This phenomenon, dubbed the "clathrate gun" (Kennett et al. 2003), is now linked by scientists not only with the PETM event but also, according to palaeontologist Peter Ward, with the Great Death of the Permian, the worst annihilation in the history of life on Earth (Ward 2008). The significance of the clathrates is that they consist of methane, a gas that is 72 times more powerful than CO_2 as a climate forcing agent in the short run, and 25 times stronger over a century or so. The clathrates could be released by a process known as 'ocean overturning', a shift in global current patterns caused by moderate warming, which brings warmer water from the surface down into the depths, to melt the deposits of frozen gas. Unlocking several trillion tonnes of methane would cause global temperatures to rocket upwards sharply. Once such a process gets under way, most experts consider, warming will happen so fast it is doubtful if humans could do anything to stop it even if they instantly ceased all burning of fossil fuels.

This 'double whammy' of global warming caused by humans releasing three trillion tonnes of fossil carbon which then precipitates an uncontrollable second phase driven by the melting of all or part of the five trillion tonnes of natural methane deposits (Buffet & Archer 2004) is the principal threat to civilisation in the twenty-first century and, combined with nuclear conflict (Chap. 4), to the survival of the human species.

The IPCC's fifth report states that the melting of between 37 and 81 % of the world's tundra permafrost is 'virtually certain' adding "There is a high risk of substantial carbon and methane emissions as a result of permafrost thawing" ((IPCC 2014a), p. 74). This could involve the venting of as much as 920 billion tonnes of carbon. However, the Panel did not venture an estimate for methane emissions from the melting of the far larger seabed clathrates and a number of scientists have publicly criticised the world's leading climate body for remaining so close-lipped about this mega-threat to human existence. The IPCC's reticence is thought to be founded on a lack of adequate scientific data to make a pronouncement with confidence—and partly to fear of the mischief which the fossil fuels lobby would make of any premature estimates. However, it critics argue, by the time we know for sure that the Arctic and seabed methane is escaping in large volumes, it will be too late to do anything about it.

The difficulty is that no-one knows how quickly the Earth will heat up, as this depends on something that *cannot* be scientifically predicted: the behaviour of the whole human species and the timeliness with which we act. Failure to abolish carbon emissions in time will make a 4–5 °C rise in temperature likely. As to what that may mean, here are some eminent opinions:

- Warming of 5 °C will mean the planet can support fewer than 1 billion people—Hans-Joachim Shellnhuber, Potsdam Institute for Climate Impact Research (Kanter 2009)
- With temperature increases of 4–7 °C billions of people will have to move and there will be very severe conflict—Nicholas Stern, London School of Economics (Kanter 2009)
- Food shortages, refugee crises, flooding of major cities and entire island nations, mass extinction of plants and animals, and a climate so drastically altered it may be dangerous for people to work or play outside during the hottest times of the year—IPCC Fifth Assessment (IPCC 2014b)
- Corn and soybean yields in the US may decrease by 63–82 %—Schlenker and Roberts, Arizona State University (Schlenker & Roberts 2009a)
- Up to 35 % of the Earth's species will be committed to extinction—Chris Thomas, University of Leeds (Thomas et al. 2004)
- Total polar melting combined with thermal expansion could involve sea levels eventually rising by 65 m (180 ft), i.e. to the 20th floor of tall buildings, drowning most of the world's coastal cities and displacing a third or more of the human population (Winkelmann et al. 2015)

- Intensified global instability, hunger, poverty and conflict. Food and water shortages, pandemic disease, disputes over refugees and resources, and destruction by natural disasters in regions across the globe—Chuck Hagel, US Secretary for Defence (Hagel 2014)
- "Almost inconceivable challenges as human society struggles to adapt… billions of people forced to relocate…. worsening tensions especially over resources… armed conflict is likely and nuclear war is possible"— Kurt Campbell, Center for Strategic and International Studies (Campell et al. 2007).
- "Unless we get control of (global warming), it will mean our extinction eventually"—Helen Berry, Canberra University (Snow & Hannam 2014).
- "2175 was a much simpler global society than now: 300 million people speaking only two major languages – English and Russian – clustered around the shores of the Arctic Ocean…" Gwynne Dyer in *Climate Wars* (Dyer 2008).

Climate of Doubt

For ordinary citizen of Planet Earth one of the most inexplicable aspects of climate change is why governments, confronted with so much reliable evidence about so grave a peril to the human future, have done so little.

The answer emerged early in 2014, when a US Senator from Rhode Island called Sheldon Whitehouse told the US Senate "I have described Congress as surrounded by a barricade of lies. Today I'll be more specific. There isn't just lying going on about climate change, there's a whole carefully built apparatus of lies. This apparatus is big, and artfully constructed: phony-baloney organizations designed to look and sound like they're real; messages honed by public relations experts to sound like they're truthful; payroll scientists whom polluters can trot out when they need them; and the whole thing big and complicated enough that when you see its parts you could be fooled into thinking it's not all the same beast" (Whitehouse 2014).

Citing a study by Robert Brulle, a professor of Sociology and Environmental Science at Drexel University, and books by Naomi Oreskes, Aaron McCright, and Riley Dunlap, Whitehouse went on to detail how, between 2003 and 2010 "140 foundations made grants totalling $558 million to 91 organizations that actively oppose climate action". These foundations were funded by the fossil fuels sector and evolved out of earlier industry campaigns to discredit the science around tobacco smoking, acid rain and ozone depletion. Since most of their funding is 'dark money', hidden from public scrutiny, the scale of their campaign worldwide is thought to be far larger. This cynical enterprise has traded on the preparedness of the media to report any claim or statement, regardless of its veracity, to manufacture the impression of a worldwide 'debate' about the validity of climate science, where none in fact exists. An attempt to scientifically analyse why politicians, in so many cases, were out-of-step with their electorates on climate change found them to be surrounded by 'echo chambers' in which the same claims came back again and again to individuals from multiple sources reinforcing their personal biases, regardless of the legitimacy of the information or its source (Jasny et al. 2015). By such methods politicians in western democracies especially have been seduced, bribed, coerced, programmed and bamboozled into not taking measures essential to the safety of their citizens and, indeed, the whole planet.

Ian Dunlop is someone who understands how this has all come about. The UK-trained engineer was for many years a senior executive with the Shell oil company and chairman of the Australian Coal Association before emerging as one of the most influential voices on the world stage about the risks of fossil fuels. As a young oil executive Dunlop was involved in modelling the future demand environment for petroleum, physical, economic and geopolitical, a role which trained him to think big-picture and long-term. When the first evidence of climate change began to amass, he says the fossil energy industry responded positively, investing in renewables and seeking ways to curb its emissions: "It was a far more progressive industry than it is today" (Dunlop I, 2015, personal communication).

What changed, he says, was a global distortion in the way the senior managers of companies are rewarded. "Since the early 1990s the bonus structure for remunerating CEOs has completely altered the corporate ethos. It has forced businesses to an intense focus on the short term, instead of planning for the longer term. Today the typical company plans three years ahead, often far less. That's no way to run a business – but it is what the market now demands."

The effect of this on coal, oil and gas firms has been to drive their executives into a quest for immediate profits—and to turn a blind eye and a deaf ear to 'long term' issues such as climate change or the fate of humanity. "These are very well-trained and educated people, technically and scientifically. They understand what is happening, and individuals will often privately agree that climate change is a huge issue for the world, for their industry and company. But then their company reward structure takes over. They've locked themselves into big personal financial commitments, and have to go along. This has produced an industry that is schizophrenic—that knows carbon is dangerous but behaves as if the opposite is true. The ethics have been sidelined.... The energy lobby is engaged in shamelessly manipulating our governments to do as little as possible, all in the name of short-term profits." A measure of their success is half-a-trillion dollars in taxpayer subsidies paid to the fossil fuel industries by governments around the world (Whitley 2013).

Dunlop says a second factor is that the fossil fuels industry is simply out of its depth, unable to cope with a circumstance it has never had to face before: the fact that humanity no longer needs it. It has proceeded on the assumption that the global system will continue to function in the twenty-first century more or less as it did throughout the twentieth century and all that was required was incremental change, or techno-fixes like carbon capture and storage (CCS). Now, however, consumers worldwide are starting to abandon fossil fuels for reasons of cost, toxic pollution and mounting alarm over climate change, and investors are spooked. Potent forces, as widely assorted as communist China (Matthews & Tan 2014) and the Roman Catholic Church (Pope 2015), capitalist America, democratic India and socialist Scandinavia, are rethinking energy and influencing global perceptions about the wisdom of oil, gas and coal. And the heavyweights are taking heed: in 2014 the world's largest mining house, BHP Billiton, conceded publicly it might have to quit coal—hitherto its largest and most profitable business—(Ker 2014) while the world's biggest coal miner, government-owned Coal India, announced a plan to plough billions of dollars into gigawatt solar power plants (Chadha 2014).

"We are facing really big risks, tipping points such as the ice sheets melting, ocean overturning, ocean acidification, the release of the methane hydrates. I doubt whether the average fossil fuel executive really understands these things – but those who do, it really scares the pants off them," says Ian Dunlop. He concludes: "I'm optimistic that humans can overcome the challenge posed by global warming – but it's going to be an extremely nasty transition."

Dollars and Sense

Back in the early nineteenth century canal owners were terrified about a new threat to their economic existence—the railways. They did all in their power to lobby governments to block and delay the advent of the iron monsters[3]. They failed—and as a result only a minuscule fraction of the world's passengers and freight go by canal nowadays, and none by horse-drawn barge. As our smart phones remind us, yesterday's technologies are constantly overtaken

---

[3] see for example, (Lewis 1959)

by better ones, more affordable and more in tune with modern demands. In the same way clean renewable energy will replace fossil fuels—the only question is whether it will be in time to make a difference to global heating… or after it is far too late.

The industry-engendered myth that renewables are costlier than fossil energy was comprehensively challenged by British economist Nicholas Stern, whose 2006 *Review on the Economics of Climate Change* concluded that unchecked global warming would cost the world at least one fifth of its economic activity and maybe up to half—whereas solving it would cost only 2 %. In other words, the cost of inaction is ten to 25 times higher than the cost of action (Stern 2006).

Furthermore, with enthusiastic support from the likes of Germany, Scandinavia, Britain, China, India, many African countries and the Vatican, it is clear that renewable energy will drive the next great phase of world economic growth. Countries which ignore this trend will be left eating the dust of early adopters. In its *2030 Market Outlook*, unashamedly-green US publisher Bloomberg predicted "By 2030, the world's power mix will have transformed: from today's system with two-thirds fossil fuels to one with over half from zero-emission energy sources. Renewables will command over 60% of the 5,579GW of new capacity and 65% of the $7.7 trillion of power investment" (Bloomberg New 2013).

Such optimism still leaves unanswered, however, the crucial question whether humanity can cut emissions in sufficient time to forestall irreversible climate change and its pan-civilisation consequences. This, alas, may ultimately be decided by the 90 large companies who own coal, oil, gas, or tar sands reserves—and whose owners appear to prize personal luxury ahead of their own grandchildren's survival.

Geoengineering Myths

Despair over the inability of billions of humans to influence the corporate self-interest of fossil fuel companies bent on increasing global carbon emissions has given rise to a number of 'technofix' proposals aimed at cooling the Earth artificially. These include:

* spraying sulphate particles into the upper atmosphere as volcanoes do, to create a sunshade effect and cool the planet
* spraying salt water above the oceans to whiten low clouds and reflect more sunlight

- cloud seeding with tiny metal flakes to reflect more sunlight
- generating microbubbles on the ocean surface to whiten it and reflect more sunlight
- thinning high cirrus clouds artificially to allow more heat to escape
- using shinier crops, trees and roofs to reflect more sunlight
- using large mirrors in space or balloons in the upper atmosphere to deflect sunlight
- feeding the oceans on iron particles to make their algae absorb more CO_2
- global reforestation, especially in the tropics and savannahs, to soak up more CO_2

All of these proposals are founded on the rather depressing assumption that humanity lacks the wisdom to prevent the fossil fuels industry from destroying the Earth's climate—and civilisation with it—and must therefore take action to offset the damage they cause.

They suffer from two major flaws. The first is that, whichever option is chosen, it will potentially disadvantage billions of people in some part of the world or other—rapid cooling, for example, would hurt those who live closer to the poles, while large-scale reforestation would affect agriculture and food production. Most climate tinkering involves increasing the scale and intensity of local floods, fires and droughts.[4] Second, it is highly doubtful if humans can control the atmosphere of a planet with the same precision as we can an infant humi-crib. It is simply too vast, too complex and has too many lags and feedbacks. Also, if we continue to burn carbon and our chosen 'air conditioning system' ever fails, the planet's temperature would spike immediately, with fatal consequences for humanity.

Many scientists now fear that individuals or groups, terrified at the growing impacts of global heating, may decide to take unilateral action to geo-engineer the planet, with incalculable and ill-considered consequences for all. Such groups could include fossil fuel companies desperate to shore-up their polluting businesses, vulnerable nations with authoritarian political systems, religious groups on a self-conceived mission to save humanity, environmental groups motivated to try to save wildlife and landscapes, and individuals or companies seeking to 'get rich quick' by marketing new global technologies. Indeed, warns Australian academic Clive Hamilton, dozens of patents have already been taken out on geoengineering technologies of highly dubious efficacy, in order to seduce gullible investors (Hamilton 2015).

---

[4] See for example, the work of the Integrated Assessment of Geoengineering Proposals (IAGP), led by the University of Leeds, UK http://www.iagp.ac.uk/.

In 2015 an expert panel convened by the US National Academy of Sciences warned against the dangers of geo-engineering, adding "There is no substitute for dramatic reductions in the emissions of CO_2 and other greenhouse gases to mitigate the negative consequences of climate change, and concurrently to reduce ocean acidification" (US National Academy of Sciences 2015).

The Secretary General of the United Nations, Ban Ki-Moon summarised the urgency of the climate situation thus: "We are the last generation that can fight climate change. We have a duty to act." (Moon 2015)

What We Must Do

1. Cease burning all fossil fuels by 2030 and replace them with renewables

   Pathway: Detailed pathways and options for climate change mitigation has been laid out by the IPCC (IPCC 2014a) and in many individual government reports. They include strategies such as replacing worn-out coal-fired power stations with clean generation, accelerated investment in renewables, carbon cap-and-trade systems, accelerated energy saving in industry, transport and cities, novel energy technologies, reafforestation and revegetation of landscapes, recycling of materials, a reduction in intensive meat production, etc.[5]

2. Reforest and replant as much of the Earth's landmass as possible (up to half)

   Pathway: this can be economically driven by carbon trading, but should follow the systematic approach outlined by EO Wilson in his book Half Earth: Our Planet's Fight for Life (Wilson 2016b) and by shifting the bulk of global food production from rural landscapes to cities (where all the necessary water and nutrients are already available to produce food with a far lower carbon and resource footprint).

3. Accelerate worldwide research and investment in clean, renewable energy

   Pathway: place renewable energy R&D on a 'war footing' worldwide, to deliver new technologies in time to avert dangerous overheating of the planet. This is a pan-species collaborative effort that has to happen at global, bilateral, national, scientific, industry and local level.

---

[5] The last applies to intensive grain-fed livestock like poultry, pigs and feedlot beef and dairy. Pastoral grazing can be managed in ways that are either neutral or which store carbon.

4. Develop global partnerships to accelerate the uptake of clean, renewable energy

   Pathway: use the model of successful government/private sector partnerships established for major infrastructure in recent decades to introduce clean energy at national, regional, urban and local community levels.

5. Shift food production from extensive agricultural systems to intensive local urban systems that use far less energy, transport, soil, water and nutrients. 'Rewild' up to half the world's existing farm land to absorb more carbon.

   Pathway: climate, water and nutrient shortages will drive this, but it can be accelerated by investment incentives and increased R&D.

6. Provide economic incentives to farmers worldwide to lock up CO_2 by 'carbon farming' and to manage 'rewilding'.

   Pathway: It will be vital to engage the support of farmers and indigenous people in their new role. See 5. above, also Chaps. 2 and 7.

7. Replace aviation and long-distance transport fuels, plastics, synthetic fibres, petrochemicals and drugs made from petroleum with renewable carbon-neutral oil made from algae and other plants. Replace urban transport with electric vehicles powered by renewable electricity.

   Pathway: governments to provide incentives for industry to move away from fossil fuels as a feedstock for transport fuels, chemicals, drugs, plastics etc in favour of natural substitutes (Chap. 7). Educate consumers to drive market demand for non-fossil products.

8. Cease all government subsidies to the fossil fuels industry immediately.

   Pathway: since so many governments and politicians receive bribes and electoral funding from fossil fuel companies to do little or nothing about climate change, this will only happen through citizen pressure, legal and media exposure of corruption in government.

9. Design cities and buildings that are carbon-neutral, save energy and recycle both nutrients and water.

   Pathway: this is a challenge for urban planners, architects, engineers and builders worldwide to conceive, plan and build the clean, green, resilient metropolises of the future. Fortunately, urban governments are far more attuned to the needs of their populations facing various existential risks etc than are most national governments and the process of competing to design renewable urban systems has begun. It needs to go faster.

What You Can Do

- Use your power as an educated consumer to send a price signal industry cannot miss, by choosing only products produced with clean energy or with declining use of fossil fuels.
- Don't vote for any politician who is not committed, deed and word, to protecting your grandchildren by defending the climate.
- Take steps to reduce your own carbon footprint—there's lots of excellent advice available.
- Educate yourself and your family about the lifecycle carbon content of manufactured goods, building materials etc. Not everything is as green as it claims.
- Consider: flying less, using public transport, walking or bicycling more, eating less meat, using fewer plastic products, using clothing to keep yourself warmer/cooler, planting more trees.
- Support companies which demonstrate a strong climate ethos.
- Favour fresh, locally-grown food. It involves far fewer transport, processing and refrigeration emissions.
- Join the rapidly growing global movement to disinvest from the fossil fuel industry and oppose new mines and extraction.
- Join neighbourhood groups who work to lower local carbon emissions or sequester carbon in the soil. Share your own knowledge and advice with others.
- Join online and global social media to share ideas, express views and send a signal to governments and corporations all around the planet that its citizens want the carbon emission ended.
- Don't invest in, work for or buy from any company that doesn't care what happens to your grandkids.

6

The Poisoner (*Homo veneficus*)

Where the people are many and their hands are all empty
Where the pellets of poison are flooding their waters…
             —Bob Dylan, *A hard rain's gonna fall, 1962*

When Marcy Borders finally died, on August 26, 2015, after a prolonged struggle against cancer of the stomach, the world media reported her as the latest victim of the terror attack on New York's twin towers in 2001, 14 years earlier. Marcy was the 28-year-old legal assistant in the smart business suit, covered head-to-toe in the grey, powdery dust released by the collapse of the two huge skyscrapers, whose heart-rending image was seared into the memories of people all over the planet in the ensuing media coverage. Somehow, she stood out as the quintessential survivor—shocked, distraught, yet still self-possessed. Years later, after the cancer made its appearance, Marcy spoke out about her experiences: "How do you go from being healthy to waking up the next day with cancer?" she asked the *Jersey Journal*. "I'm saying to myself, 'Did this thing (the towers' collapse) ignite cancer cells in me?' "I definitely believe it, because I haven't had any illnesses." (Shortell 2015). After her death, aged just 42, New York City major Bill de Blasio tweeted: "Marcy Borders' passing is a difficult reminder of the tragedy our city suffered nearly 14 years ago. NYC holds her loved ones in our hearts."

On the day of the attack on the twin towers, 9/11, a total of 2753 people perished. Over the ensuing years, thousands of the survivors in the towers and the emergency services workers who responded to them have come down with cancers and other diseases caused by toxic exposure. These are detailed in numerous scientific studies (Wu et al. 2010). In evidence before the World Trade Center

© Springer International Publishing Switzerland 2017
J. Cribb, *Surviving the 21st Century*, DOI 10.1007/978-3-319-41270-2_6

Health Program's scientific advisory committee, Laurie Breyer of the WTC Health Program stated "You can see as of May 2015, we have over 4,000 members that have been certified for a 9/11-related cancer." (World Trade Center Health Program 2015). In other words, the Twin Towers' ultimate toll from exposure to their toxic materials may far exceed that of the collapse itself.

On the other side of the world, on the same day that Marcy Borders died, more than four thousand Chinese perished from air pollution. As happens every day. In fact, scientists said, 17 % of China's population now dies from the very air they breathe (Rohde and Muller 2015). And China doesn't even have the worst air pollution in the world—that unenviable title is held by India and Pakistan (Ramsey 2015). Both Asia's urban air and the Twin Towers are poorly-understood symptoms of the increasingly toxic world we inhabit.

Planetary Poisoning

Earth, and all life on it, are being saturated with man-made chemicals in an event unlike anything which has occurred in all four billion years of our planet's story. Each moment of our lives, from conception unto death, we are exposed to thousands of substances, some deadly in even tiny doses and most of them unknown in their effects on our health and wellbeing or upon the natural world. These enter our bodies with every breath, each meal or drink, the clothes we wear, the products with which we adorn ourselves, our homes, workplaces, cars and furniture, the things we encounter every day. There is no escaping them.

Ours is a poisoned planet, its whole system infused with the substances humans deliberately or inadvertently produce in the course of extracting, making, using, burning or discarding the many marvellous products on which modern life depends[1]. This explosion in chemical use and release has all happened so rapidly that most people are blissfully unaware of its true magnitude and extent, or of the dangers it now poses to us all as well as to future generations for centuries to come.

The European Chemicals Agency (ECA) estimated in 2015 that around 144,000 different chemicals were either registered or pre-registered for use worldwide (European Chemicals Agency 2015). This is the only indication for the number of deliberately manufactured chemicals globally—and is probably therefore an underestimate. The US Government states that around 85,000 chemicals are used in that country alone and each year "an estimated 2,000 new ones are introduced for use in such everyday items as foods, personal care products, prescription drugs, household cleaners, and lawn care products" (US Department of Health and Human Services 2014).

---

[1] For a detailed description, see Cribb (2014).

The scale of the toxic risk to humans, and all life, posed by these many substances is simply unknown—but it is becoming larger by the day. World industrial chemical output is forecast to triple by the mid-century, according to the United Nations Environment Program (UNEP), in line with economic growth and with the 'chemical intensification' of the industrial economies, which occurs where more and more synthetic chemicals are used in tasks such as food production, manufacturing and processing, transportation, health care and personal adornment or as substitutes for natural products. Furthermore, the global chemical industry is rapidly migrating out of places such as Europe and North America, where it is strictly regulated, and settling in emerging industrial countries in Asia where the system is far less stringently controlled and often corrupt.

Most of the chemical substances we are exposed to daily have never been tested for human or environmental safety, warns UNEP: "Of the tens of thousands of chemicals on the market, only a fraction has been thoroughly evaluated to determine their effects on human health and the environment." In any case, it adds "Real-life exposures are rarely limited to a single chemical and very little information is available on the health and environmental effects of chemical mixtures" (United Nations Environment Program 2013).

Global output of industrial chemicals is at least 30 million tonnes a year, UNEP indicates.[2] However, these deliberately-made substances are but the tip of the iceberg of total human chemical emissions, most of which are unintentional.

Table 6.1 Estimated annual volume of contaminants emitted by human activity

• Manufactured chemicals (inc. 4 mt of pesticides)	30 million tonnes (UNEP 2013)
• Phosphorus	11 million tonnes (Rockström et al. 2009)
• Nitrogen	150 million tonnes (Rockström et al. 2009)
• Hazardous waste (including 50 mt of e-waste)	400 million tonnes (The World Counts 2015)
• Coal, oil, gas etc.	15 billion tonnes (World Coal Association 2013; US Energy Information Administration 2015; Global Energy Statistical Yearbook 2014)
• Overburden, tailings, slag and other mining wastes	>1000 billion tonnes (UNEP 2015)
• Carbon (all sources)	50 billion tonnes (IPCC 2014b)
• Materials (metals, construction, timber etc)	75 billion tonnes (OECD 2015b)
• Eroded soil	75 billion tonnes (Wilkinson and McElroy 2006)
• Water, mostly contaminated with the above	9 trillion tonnes (Hoekstra and Mekonnen 2011)

[2] A conservative estimate based on UNEP (2013), which states North American output of pollutants was 5.7 million tonnes in 2010 and the continent produced about 15%, or one sixth, of the world's total manufactured chemical output.

This is a partial list. It does not include things like nuclear and chemical weapons materials, illicit drugs, nanoparticles and so on. But it conveys a sense of the sheer volume of our emissions and their pangaean impact. Humanity's chemical emissions thus total in excess of quarter of a trillion tonnes every year and, since many of these substances are durable—such as heavy metals, persistent pesticides, plastics, carbon and sediment—a significant part of this release is cumulative, year on year. Contrary to what most people imagine, many of these substances do not break down or become safe after a time: they add to an ever-increasing, sometimes lifelong burden of toxicity for every one of us.

This chemical outpouring is, by far, humanity's greatest impact on ourselves and all life. It has only happened in the last 150 years, and especially in the last 50. Before then, none of our ancestors experienced a toxic assault on such a scale. It is arguably the most under-rated, under-investigated and poorly understood of all the existential threats that humans face in the twenty-first century.

For the first time in the Earth's history a single species—ourselves—is poisoning the entire planet.

The Great Flood

Since the 1970s science has gradually unveiled a disturbing picture of made-made chemical pollution moving relentlessly around the Earth in water, air, soil, wildlife, fish, food, trade, in people and in our very genes (see for example, Loganathan and Kwan-Sing Lam 2014).

Researchers are discovering toxic man-made substances from the stratosphere (USEPA 2014) (where it is causing a pandemic of skin cancers) to the peak of Mt Everest (Yeo and Langley-Turnbaugh 2010) where fresh snow is too polluted to drink, to the Amazon jungle (Malm 1998), to the remotest atolls (Auman et al. 1997), to the seabeds of the continental shelves (Berne et al. 1980) to the ocean deeps, where squid over a thousand fathoms down have been found to be contaminated with cancer-causing chemicals from domestic furnishings (Northeast Fisheries Science Centre 2008). From the high Arctic (Calder undated; Provieri and Pirrone 2005) to the 'pristine' isolation of the Antarctic (Konkel 2012; Australian Antarctic Division 2012; Fuoco et al. 2009). From the filthy grey-brown cloud of toxins and particulates that has settled across industrial Asia from Mecca to New Delhi to

Beijing (UNEP Centre for Clouds, Chemistry and Climate 2002) and is now spreading out to engulf the whole Northern Hemisphere and with it, most of the world's population, to the heavily polluted waters that now underlie the world's great cities (Kuroda and Fukushi 2008; van Wyck 2013; Times of India 2010) and are used for domestic drinking supplies (for example, see Environmental Working Group 2010).

Toxic man-made chemicals are routinely found by researchers in birds (Muir et al. 2002), fish (Seafish (UK) 2013), whales (Mössner and Ballschmiter 1997), seals (Ross et al. 2004), polar bears (Dietz et al. 2012) and other life-forms which have never had contact with humans, as well as throughout the global food chain (Bro-Rasmussen 1996). Scientists believe these creatures and remote regions are becoming contaminated by means of the 'grasshopper effect' in which persistent volatile organic pollutants skip around the planet in successive cycles of condensation and re-vaporisation (Semeena and Lammel 2005). Another means of moving contamination worldwide is via the mass of plastic that now contaminates the world's oceans: researchers have discovered that some durable man-made toxins can attach themselves to tiny plastic particles and so hitch a ride wherever the ocean currents take them (Rios et al. 2007).

Early in 2016, 100 national governments met in Kuala Lumpur, Malaysia, to discuss an alarming chemical development: the accelerating loss of the world's honeybees and other pollinators to global pesticide contamination. The Intergovernmental Science Policy Platform on Biodiversity and Ecosystem Services (IPBES), an independent organisation with 124 countries as members, produced the first global assessment of the damage to bees, birds, bats, beetles, moths, butterflies and other creatures which carry the pollen necessary for fertilising over three quarters of the world's main food crops and 90 % of wild flowering plants. The study found that human dependence on pollinators is increasing, and has tripled in the past half century. Out of 20,000 recognized species insect of pollinators, two in five species of insect pollinators (such as bees and butterflies) and one in six vertebrate pollinators (such as bats and hummingbirds) are on the path to extinction globally. While pesticides are a primary cause, habitat loss and climate change are also implicated. Dennis van Engelsdorp, of the University of Maryland told media "Everything falls apart if you take pollinators out of the game. If we want to say we can feed the world in 2050, pollinators are going to be part of that" (Motherboard 2016).

When Bob Dylan wrote the line "where the pellets of poison are flood-ing their waters" in *A Hard Rain's Gonna Fall* in 1962, he was prophesying a universal threat that has arisen to the world's freshwater systems from farm pesticides. In a study of surface water covering 73 countries, Sebastian Stehle and Ralf Schulz of the University Koblenz-Landau found in nearly 6000 cases worldwide the pesticide content of rivers and sediments exceeded safe levels by as much as double. "Thus, the biological integrity of global water resources is at a substantial risk," they stated. Furthermore "our results seriously chal-lenge the protectiveness of the current regulatory insecticide risk assessments and management procedures at the global scale" (Stehle and Schulz 2015).

The picture being assembled in thousands of scientific reports is of an Earth in which practically no region, including the most remote, is now free from human contaminants, which are thus directly and indirectly impairing *all life on the planet*. As we saw in Chap. 2, even widely-dispersed creatures such as frogs, honey bees, reef-building corals, beetles and deep-sea squid are affected by this global toxification—and many are consequently at risk of extinction.

Health Concerns

We humans are no exception. The evidence that we ourselves—along with our descendants, potentially for the rest of history—are at risk from the toxic flood we have unleashed is piling up in literally tens of thousands of peer-reviewed (i.e. trustworthy) scientific research reports. Despite this mass of evidence, the public in most countries is only dimly aware, or even largely unaware of what is being done to them. The reason is twofold: first, most of these reports are buried in scientific journals, written in the arcane and inac-cessible language used by specialists. The public may hear a little about certain chemical categories of concern, like pesticides and food additives, or the 'dirty dozen' (Stockholm Convention 2013a) industrial super-poisons, or 'air pol-lution' in general. However, these represent only a scant few pixels in a much larger image now amassing in the scientific literature of tens of thousands of potentially harmful substances which are disseminating worldwide. Second, the proportion of chemicals which have been well-tested for human safety is quite small, and next to nothing is known about their toxicity when they interact with other substances, both natural and synthetic, in our daily living environment, food or bodies.

Tests around the world reveal that the average person nowadays is a walking contaminated site: citizens of an advanced societies may carry several hun-dred industrial chemicals in their body tissues, blood or bones at any one time (Thornton et al. 2002; Onstot et al. 2010; Ruiz 2010). The US Centers for Disease Control, for example, conducts a regular survey of 212 noxious substances, finding several of them in the blood or urine of almost all the Americans it tests, and many of them in a majority (US CDC 2014).

The US Environmental Working Group (EWG), in independent tests, identified 414 industrial toxins in 186 people ranging in age from newborns to grandparents. It also found 212 industrial chemicals, including dioxins, flame retardants and known carcinogens in the blood of new-born babies of minority ethnic groups, who had suffered contamination while still in the womb (Environmental Working Group 2009). These findings have raised such concern among medical scientists internationally that they embarked on the longitudinal testing of almost three quarters of a million infants in seven countries. The aim is to monitor the child's chemical body burden and try to define how it does - or doesn't - relate to any health problems they may later suffer. Running for as long as 20 years, clear results are not expected until the mid-2020s.

Today's newborn has barely taken her or his first breath than the poisoning continues. The World Health Organisation reports that mother's milk was found to be contaminated with up to 22 noxious pesticides and industrial chemicals in over 70 countries worldwide including America, 15 European nations, Brazil, China, Russia, India, Australia and numerous African and Asian countries (Stockholm Convention 2013b). The contamination of mother's milk with industrial chemicals has been reported in the scientific literature since the 1950s, but its incidence has continued to rise both globally and in individual countries well into the twenty-first century, with little sign of abating: where testing shows levels of one chemical (such as the banned pesticide DDT) may be in decline, many new ones (like flame retardants and 'preservatives') are emerging to replace it. Indeed, the incidence of poisoned mother's milk is now used by some health authorities as an indicator for the level of poisoning in the population as a whole. This gives some notion of how intractable a problem it is to protect the twenty-first century child from man-made toxins. For those who might resort to infant formula instead of breast milk, the news that all American commercial infant milk powders tested were found to be contaminated with rocket fuel (Sharp 2009), while China executed two milk company executives for adding the industrial plastic melamine to powdered baby milk (Foster 2011), is not reassuring. The issue to bear in mind is that, for more than a million years, babies came into the world and were nurtured and raised without exposure to industrial toxins. Since the 1940s this has all changed, radically. For practically every baby born on Earth today, its first drink contains man-made chemicals.

The situation is not improving. The Californian Government, for example, states "More and more scientists and toxicologists are identifying 'emerging chemicals of concern,' or ECCs." Recent studies have shown that some of these chemicals can act as endocrine disruptors, which means they block or distort the normal signals our hormones send to our body, and can produce these effects in vanishingly small amounts, at parts per billion or per trillion level. Also, the effects of ECCs can be transgenerational: scientists have found that when laboratory animals are exposed as embryos, the effects are transmitted not only to them, but also to their offspring for several generations thereafter (Manikkam et al. 2013). In addition, scientists are concerned about the unknown effects of exposures to mixtures of these ECCs and/or other chemicals (California Department of Toxic Substances Control, Emerging Chemicals of Concern 2007).

The modern industrial food supply is a major daily source of toxins for most people. According the US Environment Protection Authority some 400,000 tonnes of pesticides are used to grow America's food each year. Since America uses about 22 % of the world's pesticides, this puts global use of specialised poisons in farming alone at around 1.8 million tonnes (US EPA 2013). That these are getting into consumers is clearly proven by the worldwide incidence of pesticides found in mother's milk as well as the blood of infants. However, in addition to around 2000 chemicals used to grow food on the farm at least 4000 further chemicals are then used during food storage and processing, as additives, preservatives or in the materials and containers in which food is packaged (Muncke et al. 2014). Scientists have expressed alarm that many of these substances, singly or in combination with one another, may prove harmful to human health over the long term (ScienceDaily 2014). And while most parents would be shocked to learn they are feeding their children on fossil fuels, many of the brightly coloured dyes used in the food manufacturing industry are in fact derived from coal tar or petroleum and have been linked with conditions ranging from cancer and brain damage to reproductive disorders (Kobylewski and Jacobson 2010). Human exposure to food-chain chemicals is presently almost unavoidable, lifelong: even a diet of home-grown food is susceptible to air, water and soil pollution, while wild animals and fish are now extensively contaminated.

One of the first things a modern teenage girl does is apply cosmetics and 'body care' products. Unwittingly she thus exposes her developing reproductive system to a host of chemicals, some of which are now linked with hormonal and reproductive disorders, even infertility and gender changes. The University of Toronto, which conducts one of the world's leading research programs into cosmetics, states "Exposure to perfumes and other scented products can trigger serious health reactions in individuals with asthma, allergies, migraines, or chemical sensitivities. Common scented products include perfume, cologne, aftershave, deodorant, soap, shampoo, hairspray, bodyspray, makeup and powders… air fresheners, fabric softeners, laundry detergents, cleaners, carpet deodorizers, facial tissues, and candles (University of Toronto 2013).

"Fragrance chemicals are, by their very nature, shared. The chemicals vaporize into the air and are easily inhaled by those around us. Today's scented products are made up of a complex mixture of chemicals which can contribute to indoor air quality problems and cause health problems," it says. In particular, the use by mothers of volatile fragrances may affect their unborn or newly-born child at a particularly sensitive stage of its development. Like chemicals in food, fragrances are almost unavoidable for the majority of urban citizens—like tobacco smoke and other forms of air pollution, they are all around us.

That the poisoning of people is a now a whole-of-life issue, and beyond, as confirmed by an interesting piece of Australian research which found that, even when dead and buried, we release our body burden of durable toxins and heavy metals accumulated over a lifetime back into the groundwater that flows beneath our major cities—and which is often used for drinking (Dent 2002). Our poisons thus live on, to affect others, even when we ourselves are gone.

The Risks

"Exposure to toxic chemicals can cause or contribute to a broad range of health outcomes. These include eye, skin, and respiratory irritation; damage to organs such as the brain, lungs, liver or kidneys; damage to the immune, respiratory, cardiovascular, nervous, reproductive or endocrine systems; and birth defects and chronic diseases, such as cancer, asthma, or diabetes. The vulnerability and effects of exposure are much greater for children, pregnant women and other vulnerable groups," says the UN Environment Program (UNEP 2013).

UNEP adds that about five million people die and 86 million are disabled yearly by chemicals directly, making them one of the world's main causes of death—many times more lethal than malaria, HIV, ebola or tuberculosis, for example. However, this number massively understates the true toll, as both the WHO and UNEP concede. It does not, for example, include millions more deaths and disabilities where chemicals are implicated in cancers, heart disease, obesity, diabetes and mental disorders. Nor does it include the slow damage which a lifelong personal burden of toxics may cause. A particular concern is how chemicals—intentional and unintentional—interact with the thousands of other compounds in our environment and daily intake to form billions of potentially toxic mixtures, most of them unknown and un-investigated.

The World Health Organisation has estimated that 12.6 million people die yearly from living or working in an unhealthy environment—almost one in four of all deaths worldwide. These deaths are attributable to "…air, water and soil pollution, chemical exposures, climate change, and ultraviolet radiation" it stated in a report release in 2016. Furthermore, most of those deaths were considered preventable (WHO 2016).

One of the most appalling dimensions of the global chemical flood is the harm it is doing to children, without their having any say in the matter: among the deaths noted by WHO, at least 1.7 million were children. Commenting on just one aspect of this harm—brain damage—Harvard professors Philippe Grandjean and Philip Landrigan wrote in The Lancet in 2014:

Neurodevelopmental disabilities, including autism, attention-deficit hyper-activity disorder, dyslexia, and other cognitive impairments, affect millions of children worldwide, and some diagnoses seem to be increasing in frequency. Industrial chemicals that injure the developing brain are among the known causes for this rise in prevalence. In 2006, we did a systematic review and identified five industrial chemicals as developmental neurotoxicants: lead, methylmercury, polychlorinated biphenyls, arsenic, and toluene. Since 2006, epidemiological studies have documented six additional developmental neuro-toxicants—manganese, fluoride, chlorpyrifos, dichlorodiphenyltrichloroethane, tetrachloroethylene, and the polybrominated diphenyl ethers. We postulate that even more neurotoxicants remain undiscovered. To control the pandemic of developmental neurotoxicity, we propose a global prevention strategy. Untested chemicals should not be presumed to be safe to brain development, and chemi-cals in existing use and all new chemicals must therefore be tested for develop-mental neurotoxicity (Grandjean and Landrigan 2014).

Brain impairment may affect many people in modern society: recent stud-ies by Harvard School of Public Health and Lawrence Berkeley National Laboratory found that people in buildings exposed to 400 ppm of carbon dioxide in the air they breathed, as well as volatile organic compounds (from plastics, furnishings etc) had significantly lower intelligence scores and poorer decision-making abilities than those breathing clean air. The Harvard study found people with clean air had cognitive scores 61–101% higher! (Allen et al. 2015). These studies applied to polluted indoor air, but with CO_2 levels now above 400 ppm in the open air, they foreshadow a risk that the atmosphere itself is becoming adverse to human intelligence.

Medical research is reporting unexplained increases in once-rare conditions whose modern upsurge scientists are now increasingly linking, in part, to humanity's continual multiple chemical exposure. According to David Carpenter of New York State University, these mixtures can prove far worse than single chemicals in their impact on the human body (Carpenter et al. 2002a). Their effects include developmental disorders, sexual dysfunction (including infertility), nerve, brain and bone diseases (including conditions like autism, depression, Parkinson's and Alzheimer's), cancers and heart disease. A large international study of 85 suspect chemicals found that even low doses of comparatively innocuous chemicals may trigger cancers when combined with one another in the diet or living environment (Goodson et al. 2015). The scientists called for a complete overhaul of chemical regulation on cumulative exposure of humans to possible carcinogens.

There is also mounting evidence connecting the global obesity pandemic with the effects of endocrine disrupting chemicals distorting the body's energy storage and distribution systems (Porta and Lee 2012). The World Health Organisation and UNEP, in a survey of recent science, warned that reproductive and other hormonal disorders are on the rise globally, that man-made substances are increasingly implicated by laboratory studies, and that the scale of the problem is probably underestimated (WHO 2012). Falling sperm counts in males, reduced fertility in females, genital deformities, changes in gender and sexual orientation are all now linked to endocrine disrupting chemicals. As mentioned earlier, there is growing evidence that the effects of a parent being poisoned may even be passed on to their children and grandchildren for several generations, by epigenetic means (Hou et al. 2011).

Furthermore, the toxic flood may be connected with a range of recent and previously unknown diseases and conditions, among them Multiple Chemical Sensitivity (MCS), Gulf War Syndrome, Stiff Person Syndrome, Irritable Bowel Syndrome (IBS), Cycle Vomiting Syndrome (CVS), Electromagnetic Hypersensitivity, Chronic Fatigue Syndrome (CFS), and Attention Deficit Hyperactivity Disorder (ADHD). These are conditions for which doctors have, as yet, no clear explanation—if indeed they are diseases. However, the situation has not been helped by a tendency on the part of some doctors to 'blame the patient' by attributing them to a mental or genetic condition, rather than actively seeking a cause (Kirmayer et al. 2004).

There are two essential points about the Earthwide chemical flood. First it is quite new. It began with the industrial revolution of the late nineteenth century, but expanded dramatically in the wake of the two world wars—

where chemicals were extensively used in munitions—and has exploded in deadly earnest in the past 50 years, attaining a new crescendo in the early twenty-first century. It is something our ancestors never faced—and to which we, in consequence, lack any protective adaptations which might otherwise have evolved due to constant exposure to poisons.

Second, the toxic flood is, for the most part, preventable. It is not compulsory—but is an unwanted by-product of economic growth. Though driven by powerful industries and interests, it still lies within the powers and rights of citizens, consumers and their governments to demand it be curtailed or ended and to encourage industry to safer, healthier products and production systems.

The issue is whether, or not, a wise humanity would choose to continue poisoning our children, ourselves and our world.

Regulatory Failure

Despite the fact that around 2000 new chemicals are released onto world markets annually, most have not received proper health, safety or environmental screening—especially in terms of their impact on babies and small children. Regulation has so far failed to make any serious curtailment of this flood: only 21 out of 144,000 known chemicals have been banned internationally, and this has not eliminated their use. At such a rate of progress it will take us more than 50,000 years to identify and prohibit or restrict all the chemicals which do us harm. Even then, bans will only apply in a handful of well-regulated countries, and will not protect the Earth system nor humanity at large. Clearly, national regulation holds few answers to what is now an out-of-control global problem.

Furthermore, the chemical industry is relocating from the developed world (where it is quite well regulated and observes its own ethical standards) and into developing countries, mainly in Asia, where it is largely beyond the reach of either ethics or the law. However, its toxic emissions return to citizens in well-regulated countries via wind, water, food, wildlife, consumer goods, industrial products and people. The bottom line is that it doesn't matter how good your country's regulations are: you and your family are still exposed to a growing global flood of toxins from which even a careful diet and sensible consumer choices cannot fully protect you.

The wake-up call to the world about the risks of chemical contamination was issued by American biologist Rachel Carson when she published *Silent Spring* in 1962, in which she warned specifically about the impact of certain persistent pesticides used in agriculture. Since her book came out, the volume of pesticide use worldwide has increased 30-fold, to around four million tonnes a year in the mid-2010s. Since the modern chemical age began there has been a string of high-profile chemical disasters: Minamata, the Love Canal, Seveso, Bhopal, Flixborough, Oppau, Toulouse, Hinkley, Texas City, Jilin, Tianjin. Most of these display a familiar pattern of unproductive confrontation between angry citizens, industry and regulators, involving drawn-out legal battles that deliver justice to nobody. By their spectacular and local nature, such events serve to distract from the far larger, more insidious and ubiquitous, universal toxic flood.

Chemists and chemical makers often claim that their products are 'safe' because individual exposure (e.g. in a given product, like a serve of food) is too low to result in a toxic dose, a theory first put forward by the mediaeval scholar Paracelsus in the sixteenth century. This 'dose related' argument is disingenuous, if not dishonest—as modern chemists well know—for the following reasons:

- Most chemicals target a receptor or receptors on certain of your body cells, to cause harm. There may be not one, but hundreds or even thousands of different chemicals all targeting the same receptor, so a particular substance may contribute an unknowable fraction to an overall toxic dose. That does not make it 'safe'.
- Chemicals not known to be poisonous in small doses on their own can combine with other substances in water, air, food or your body to create a toxin. No manufacturer can truthfully assert this will not happen to their products.
- Chemical toxicity is a function of both dose and the length of time you are exposed to it. In the case of persistent chemicals and heavy metals, this exposure may occur over days, months, years, even a lifetime in some cases. Tiny doses may thus accumulate into toxic ones.
- Most chemical toxicity is still measured on the basis of an exposed adult male. Babies and children being smaller and using much more water, food and air for their bodyweight, are therefore more at risk of receiving a poisonous dose than are adults.

Chemicals and minerals are valuable and extremely useful. They do great good, save many lives and much money. No-one is suggesting they should all be banned. But their value may be for nothing if the current uncontrolled, unmonitored, unregulated and unconscionable mass release and planetary saturation continues.

Chemical Extinction

Two billion years ago, excessive production of one particular poisonous chemical by the inhabitants of Earth caused a colossal die-off and threatened the extermination of all life. That chemical was oxygen and it was excreted by the blue-green algae which then dominated the planet, as part of their photosynthetic processes. After several hundred million of years, the planet's physical ability to soak up the surplus O_2 in iron formations, oceans and sediments had reached saturation and the gas began to poison the existing life. This event was known as the 'oxygen holocaust', and is probably the nearest life on Earth has ever come to complete disaster before the present (Margulis and Sagan 1986). Since it developed slowly, over tens of millions of years, the poisonous atmosphere permitted some of these primitive organisms to evolve a tolerance to O_2—and this in time led to the rise of oxygen-dependent species such as fish, mammals and eventually, us. The takehome learning from this brush with total annihilation is that it is possible for living creatures to pollute themselves into oblivion, if they don't take care to avoid it or rapidly adapt to the new, toxic environment. It's a message that humans, with our colossal planetary chemical impact, would do well to ponder.

While it is unlikely that human chemical emissions alone could reach such a volume and toxic state as to directly threaten our entire species with extinction (other than through carbon emissions in a runaway global warming event) or even the collapse of civilisation, it is likely they will emerge as a serious contributing factor during the twenty-first century in combination with other factors such as war, climate change, pandemic disease and ecosystem breakdown. Credible ways in which man-made chemicals might imperil the human future include:

- Undermining the immune systems, physical and mental health of the population through growing exposure to toxins
- Reducing the intelligence of current and future generations through the action of nerve poisons on the developing brains and central nervous systems of children, rendering humanity less able to solve its problems and adapt to major changes; and by increasing the level of violent crime and conflict in society, which is closely linked to lower IQ.
- Bringing down the economy through the massive healthcare costs of having to nurse, treat and maintain a growing proportion of the population disabled by lifelong chronic chemical exposure.

- By poisoning the ecosystem services—clean air, water, soil, plants, insects and wildlife—on which humanity depends for its own survival and thereby contributing to potential global ecosystem breakdown
- By augmenting the global arsenal of weapons of mass destruction and hence the risk of their use by nations or uncontrollable fanatics.

Who Is Responsible?

Solving the threat posed by chemical saturation to the human future starts with recognising that we, ourselves, are responsible. It is we who generate the economic demand which leads to the mass production and universal release of toxins.

Every act of consumption on a crowded planet has chemical consequences. Those consequences now claim or damage tens of millions of lives. So, in a sense, *we are all getting away with murder.*

Acknowledging this uncomfortable truth is essential to modern society taking action to clean up the Earth and to protect our children into the future. If we have given rise to the problem by demanding goods which are toxic or made with toxic processes, then we alone have the power to correct it. It is clear that governments collectively lack both the capacity and the will to regulate the global toxic flood, and that industry will not be motivated to change its ways without a clear market signal from us. Regulation is important - but if we rely on rules alone to protect our children the evidence so far indicates they will not succeed.

In a globalised world only we, the people, are powerful enough, as consumers, to send the market signals to industry to cease its toxic emissions. And especially, to *reward* it for producing clean, safe, healthy products or services that we value. This is becoming possible on a global scale thanks to the universal penetration of the internet. For the first time in human history, we have the means to share a universal understanding of a common threat, to educate and learn from each other about what we can all do to mitigate it. Through the internet and social media it is possible to mobilise world awareness among citizens, across borders, cultural, ethnic, religious and economic divides, to come together to cleanse our poisoned planet.

If we undertake this, it will be an expression of people power and genuine global democracy like none before in history. An undeniable expression of the will and wisdom of humankind and, potentially, the first step towards our 'thinking as a species' (Chap. 10). Furthermore, it is already happening: across the world millions of concerned citizens, parents and consumers are already joining minds at lightspeed on the internet, to share knowledge, advice and ideas about how to protect our children and clean up society. The leading ten organisations which represent concerned citizens or give them a voice had a total membership of 50 million in 2014 (Cribb 2014, pp. 233–235) and all are growing rapidly. It they act together on common issues, such as toxic exposure, their sheer numbers form the basis for a uniquely powerful citizens' movement, spanning the entire planet and bridging all the main boundaries that divide us.

Many Solutions

A host of promising new technologies are available to clean up our planet. These range from 'green chemistry'—developing softer, less toxic chemicals—to zero waste, industrial ecology, green manufacturing and building, renewable energy, organic agriculture, integrated pest management, product stewardship, recycling, life cycle assessment, risk assessment and remediation (see Cribb 2014, pp. 195–200) … the list is both long and heartening. All that are lacking are the will—and the economic incentives.

Furthermore, the main toxic threat can also be quite practically reduced— by banning or phasing out the use of coal, oil or natural gas, which are the primary sources of most man-made toxins. Since this is also the solution to global warming, it follows that there are now two unarguable existential reasons for humanity to transition to safer, cleaner and more sustainable sources of energy, chemicals and materials. The worldwide movement to divest in coal and other polluting fuels offers a practical pathway for achieving this.[3]

Far from being disadvantageous to industry, a universal demand for clean products will open fresh markets, generate more jobs, grow new industries and companies and engender greater health, safety and prosperity for the community as a whole—just as renewables are now powering the global energy revolution. But it will call for a wiser humanity to advocate and provide the economic signals that will drive it.

---

[3] 350.org. 2016. About Fossil Free. http://gofossilfree.org/about-fossil-free/

A Human Right

Every person in the world has a right to life, liberty, personal security, marriage and family, equality, work, education, to freedom of belief, to freedom from torture. These are available to each of us under the Universal Declaration on Human Rights (UDHR 2016).

It is therefore more than a little disturbing to find there is no human right not to be poisoned. Not for us, nor for our children.

A child born today may enjoy many of the rights listed under the Universal Declaration—but not the right to a full intelligence, to undamaged genes, to a life free of cancer, mental or sexual dysfunction or other disorders increasingly linked by science to chronic lifelong chemical exposure. A child born today does not even enjoy the same freedom from toxic exposure as its own ancestors. The lack of such a right, in the presence of a right to leisure, to social security or to cultural participation bespeaks an irrational blind spot in the contemporary conscience, deriving either from an acute lack of awareness of the scale of the problem or from a desire not to hear news which is distasteful or disturbing.

Article 5 of the Universal Declaration proclaims that every human has a right not to be tortured. Although this is a right which, presumably, applies to only a small percentage of the world population at any one time, there is no similar right for people to be safe from the flood of toxins or suspected toxins that now inundates virtually the entire human race, cradle to grave. To be safe from an assault that kills millions and harms tens of millions more—often in ways that might well be regarded as torture if you caught someone deliberately inflicting them, is a serious oversight in the UDHR. Some may regard a Right Not to Be Poisoned as unnecessary, or else covered by other rights such as the Right to Life. However, such a Right is necessary in order to implant in the human consciousness a greater awareness of the universal scale of the damage being inflicted by this particular transgression of our rights—and as a means of bringing industry and government to face their responsibilities and consumers to send the essential economic signals in favour of cleaner, safer, healthier products and processes.

Until, and unless, we have a Right *Not to be Poisoned,* there will never again be a day in our history when we are not.

What We Must Do

1. Form a global network of people and institutions to Clean Up the Earth.
   Pathway: most of these organisations are on the internet and many are now combining in real time to share information and pressure governments for a cleaner world. They need our encouragement, support and participation to go faster.

2. Spread awareness, share knowledge and motivate industry to adopt clean production systems and help citizens to become 'clean consumers'.
   Pathway: this is already taking place via the internet and social media, but needs to accelerate and extend into schools and to consumer education.

3. Implement a universal human right not to be poisoned.
   Pathway: Since everyone is being poisoned it follows there is a universal need for such a right. Consumer groups, lawyers and human rights bodies can make the case for its inclusion in the UDHR.

4. Replace all coal, oil and other fossil fuels with clean energy and with clean feedstocks for industry.
   Pathway: see Chap. 4.

5. Eliminate use of all known toxic substances from the food chain, water supplies, personal care products, home goods and the wider environment.
   Pathway: Stronger citizen pressure is required to compel governments to act on known toxins. Most of them know what is bad for us but, under industry pressure, do little to prevent it. Citizens need to become informed consumers, shun all products containing or releasing known toxins and expose weak regulation that endangers lives.

6. Press for prevention of disease in medicine, as opposed to chemical 'cures' for diseases often caused by chemicals.
   Pathway: citizen pressure is needed to refocus the medical system on the needs of consumers for disease prevention, rather than those of pharmaceutical companies which prefer diseases to be long and costly to treat.

7. Train all young chemists, scientists and engineers in their social and ethical responsibility to 'first, do no harm'.

    Pathway: all major scientific disciplines should introduce an oath or pledge by graduates not to use their science in any act or technology that may cause harm to humanity. This is already part of the Hippocratic Oath taken by medical doctors on graduation from university. There being no ethical reason for universities to train other science graduates to harm humans, a universal Hippocratic injunction is desirable to inculcate harm prevention in all professional codes from the start.[4]

8. Empower industry to make profits ethically, by producing clean products that do no harm.

    Pathway: the quickest way to clean up industry is for consumers to send it the economic signals that they value clean products—and the quickest route to that is to educate consumers about what is safe and what is toxic. Many organisations are already doing this, but the learning needs to go farther, faster and wider, via the internet.

9. Reward industries which adopt approaches such as green chemistry, product stewardship and zero waste with our patronage and support.

    Pathway: improve consumer education about what is safe and what isn't.

10. Implement mandatory toxicity testing of all new industrial substances and major waste streams.

    Pathway: this requires both national and global regulation, led by bodies such as the Stockholm Convention and reinforced by government environmental protection agencies, parents' and consumer bodies.

---

[4] This was first proposed by Nobel Laureate Josef Rotblat in his acceptance speech in 1995. https://en.wikipedia.org/wiki/Hippocratic_Oath_for_scientists

What You Can Do

- Learn to distinguish between consumer products that contain toxins or are made with toxic processes: exclude them from your home, work and life.
- Above all, keep them away from your kids. Don't feed children on foods containing substances made from fossil fuels (e.g. colourings or pesticides).
- Use your power as a consumer to shun companies that emit toxins—and reward those which produce clean, safe products and use clean processes
- Participate actively in parents', citizens' and consumer groups dedicated to cleaning up the Earth, or just your local community.
- Educate your children to choose wisely among products and services, based on their personal and universal health impact. Empower children to educate us.
- Use your power as a voter to compel governments to take their duty of care towards children and future generations more seriously and to strengthen regulation and oversight of all chemical emissions, deliberate or not.

7

The Devourer (*Homo devorans*)

Lo que separa la civilizacion de la anarquia son solo siete comidas.
(There are only seven meals between civilisation and anarchy)
—*Spanish proverb.*

Touch your chin and reflect upon the following: the human jawbone is among the most destructive of implements on the planet. For every meal that each of us consumes, 10 kilos of topsoil are lost and 800 litres of fresh water expended, while 1.3 l of diesel and a third of a gramme of pesticide are used and 3.5 kilos of CO_2 are released. And that's just one meal, for one person.

Using your calculator, multiply each of these numbers by 1100 for approximately all the meals you eat in a year and then multiply the result by 10,000,000,000, which is the projected human population in the 2060s. In just a few seconds, you will begin to see why our modern food system is in trouble and wild animals are going extinct. So great have our appetites and numbers become that we are now devouring the Earth—and eating your own planet is not an act concordant with wisdom.

Modern food systems depend on technology—but they also depend on finite resources such as soil, oil, fish, fertiliser, and clean water: put simply, the dilemma we face in the twenty-first Century is that we have an expanding global demand for food that is reliant on a contracting global resource base. Furthermore, a food system that depends critically on good weather is likely to suffer in a world becoming more climatically volatile. For these reasons both the diet and how we produce our food must, and will, change. Indeed, over the twenty-first century food will evolve more rapidly and amazingly than in any previous era of history. We are, in fact, embarking on one of the great ages of humanity, *The Age of Food*.

© Springer International Publishing Switzerland 2017
J. Cribb, *Surviving the 21st Century*, DOI 10.1007/978-3-319-41270-2_7

What humans eat a century from now, how we produce and consume it, its health value and composition will seem as strange to us today, as our modern foods might appear to our own great great grandparents—before the age of cosmopolitan cuisine, cold storage, takeaway, manufactured food, celebrity chefs and cooking shows. This food revolution will arise as a result of implacable demand and resource pressures building up throughout the global food system, coupled with the advent of remarkable new technologies and emerging popular trends in farming, health and sustainability.

Food Insecurity

There are ten main factors which drive global food insecurity, two on the demand side and eight on the supply side. Most attempts to explain what is happening in world food tend to overlook several of these factors, yet all are of importance, all are interwoven and all must be addressed together if the global food supply is to remain secure through the twenty-first Century, the era of 'peak people'.

On the demand side, the requirement for a doubling in global food production is driven by population growth (which is cased both by birth rates *and* by people living longer lives) and rising living standards coupled with economic demand for higher quality, richer, more nutritious foods, especially in developing and newly industrialising countries.

On the supply side, the main things that limit our ability to double food production are:

- Physical loss and decline in fertility of soils worldwide, combined with a shrinking world farming area
- Scarcities of fresh, clean water in heavily populated regions while megacities and the energy sector combine to steal farmers' water, reducing the amount available for food production
- Uncertain availability and high cost of liquid transport fuels out to mid-century and beyond
- Emerging scarcities of high-quality mineral fertilisers by mid-century, making them unaffordable to most farmers
- Continuing decline and potential collapse of wild fish stocks due to over-fishing and ocean pollution
- Global decline in public sector investment in food, agricultural and fisheries science, leading to stagnation in crop, pasture and fisheries yields and a marked increase in diet-related disease across the human population

- A worldwide drought of 'patient capital' for new investment in farming and food production, along with speculative investment in farm land and commodities and 'landgrabs' by speculators and rich corporations
- Extinction of the temperate climate which gave rise to agriculture and its replacement by a far more uncertain climatic regime characterised by more frequent and intense floods, storms and droughts, heatwaves and loss of production on farms.

It is the synergy between these ten drivers that is the primary cause of global food insecurity, present and future. Solving several of them does not solve the food problem: a food secure world requires the solution of *all ten* problems simultaneously, and in ways that do not conflict with one another. There are no miracle cures: even the highest-yielding 'super crop' still relies on having adequate soil, water, oil, chemicals and fertiliser and a stable climate to grow in. The sheer scale of the challenge of raising output at a time of global resource contraction is poorly grasped by governments, consumers and much of the food industry itself.

The Spanish, who undoubtedly learned it from bitter experience, have an old saying that "*There are only seven meals between civilisation and anarchy*": if the population goes unfed for more than a couple of days, heads at the top are liable to roll. Historically, both the French and Russian revolutions arose out of famines, the genocide in Rwanda and the civil war in Darfur both originated in disputes over food, land and water between ethnic groups, and in 2012 governments in Egypt and Tunisia fell as a result of popular movements that began as food protests. In the twenty-first Century regional famines, hunger and disputes over food, land and water are potential triggers for civil insurrection, government collapse, refugee tsunamis, genocides and even international wars.

How Many People?

Tonight around two hundred thousand more people will sit down to dinner than dined last night. Growth in the human population, however, is not simply a matter of babies born: it is also down to well-off people living longer lives. The average resident of a country with a high life expectancy will consume 35,000 more meals than someone living in a country where most people still die comparatively young (World Health Organization 2012). Furthermore, a baby born to an affluent couple will occupy from six to ten times more of the planet's limited resources than does one born to poor parents. Rich societies thus contribute disproportionately to the stress on the world's increasingly limited resources of soil, water, minerals and energy—and hence to the risk of famine and conflict over those resources. It follows that, if all humans are to enjoy a moderate to good standard of living, the planet can only support a fraction of its present population over the long-run.

By the close of this century—barring major crises—there could well be 11 billion mouths to feed. The United Nations Population Division's medium fertility forecast estimates the world population will hit 10 billion in 2062 and 11 billion in the 2090s (UN ESA 2014). Some commentators, such as Professor of Statistics Adrian Raftery of Washington University, caution it could go as high as 12 billion (Hickey 2014). On the other hand, argues Paul Ehrlich, author of the famous 1969 book *The Population Bomb,* the global spread of feminism and growth in women's economic opportunities combined with a campaign to encourage smaller families, could equally see the human population peak at 8.5 billion in the 2050s, and then commence a slow decline (Ehrlich and Ehrlich 2014). The Earth's long-term carrying capacity is uncertain, depending as it does on the living standards and food choices of individuals: biologist E.O. Wilson put it at ten billion—provided, he said, everyone turned vegetarian (Wolchover 2011). However, as noted in Chap. 3, humans are already using 1.6 Planet Earths to meet our current needs—so the limiting factors may well be resources like soil and water (World Footprint Network 2016). This implies the planet's *long-term* carrying capacity of humans, living at a reasonable standard, may be closer to the four billion it held in 1974. Some, such as Ehrlich, argue it is more like two billion, at our present living standards. These are the sorts of targets a wise humanity would be aiming for over the coming century if it wishes to ensure its long term survival, wellbeing and prosperity. In the meantime, however, we have to plan to feed the world through a period of several decades of 'peak people' until the women of the world can sufficiently lower human numbers by reducing their own fertility, as they are presently doing in all regions of the globe.

Predictions about the size of food demand required to satisfy the world population out to 2050 range from a growth rate of 1.1–1.5 % every year (McKenzie and Williams 2015). This puts growth in food demand at between 59 and 98 % in the coming 40 years. The large variation depends on whether one assumes newly-affluent countries will adopt western levels of meat-eating—already evident in China, for example—or stick to a predominantly vegetarian diet (as many Indians may prefer).

To meet both population *and* economic demand growth in places like China, India, Latin America, Africa and the Middle East and beyond, global food availability from all sources must double within half a century. In volume terms, this involves producing more food in the coming 40 years than we did in the last 5000 (The Economist 2015a). However almost everything needed to do this by traditional agriculture is becoming scarce: soil, water, nutrients, energy, technology, fish, capital and a stable, reliable climate.

It is this collision between burgeoning demand and shrinking resources that makes food the primary challenge of our age—more immediate, even, than climate change (Cribb 2011).

Water Wars

The first great scarcity to strike agriculture will be water. By the 2020s about 2.9 billion people in 48 countries will be by facing with acute water stress, a United Nations report warns. "The need to grow more food, produce more energy, and increase luxury goods production will drive ever greater demand for water... The ultimate consequence is that, by 2030, demand for water could be 40% greater than supply available." (Schuster-Wallace and Sandford 2015b)

The average citizen of Planet Earth uses about 1386 tonnes of fresh water a year—a thousand tonnes of it in the form of food (Hoekstra and Chapagain 2007). In the course of a lifetime, we each use enough fresh water to float the *USS Enterprise*, a rather large aircraft carrier (95,000 tonnes). That refreshing morning cup of coffee took 140 l of water just to grow the beans, and your slice of toast 40 l for the wheat. That neat little T-shirt took 4 tonnes of water merely to produce the cotton. Your evening glass of wine involved 120,000 ml of water just to produce 150 ml of fermented grape juice. Your chicken dinner required 6 tonnes of water to grow the grain that fed the chook that you ate (Lenntech 2014). If it was red meat, a kilo of grainfed beef uses around 15 tonnes of water. Yet few of us spare a second's thought for the prodigious volumes of water embodied in our daily diet—or the colossal impact that it's withdrawal from rivers and aquifers is having on fresh water supplies and on landscapes worldwide. Around 70% of all the fresh water used by humanity goes into production of irrigated crops and pastures (UN Food and Agriculture Organisation 2014) which in turn supply nearly half the world's food (the rest is from rain-fed farming). And the share of irrigated food production is growing year by year, as water supplies in dryland farming regions become increasingly stressed by climate change. As British author Fred Pearce graphically put it "as a typical meat-eating, beer-swilling, milk-guzzling Westerner, I consume as much as a hundred times my own weight in water every day." (Pearce 2006)

This over-extraction of fresh water for food production is having a punitive impact on the world's rivers, lakes and groundwater sources, especially in warm, dry regions. In many of the world's food bowl regions groundwater is being extracted to water crops far faster than it recharges, with water tables in parts of China and India diminishing by a metre or more a year (Doll et al. 2012). As we noted in Chap. 3, groundwater is important because it constitutes 95 % or more of the available fresh water on the planet, and once removed, is only replaced at geological rates: for example, it is estimated that once pumped dry, parts of the Ogallala aquifer that waters farms on the American High Plains may take up to 6000 years to refill (Biello 2012). Groundwater currently supplies more than 40 % of the world's food (Wada 2012) and by the mid-century, according to UN Food and Agriculture Organisation estimates, it may need to supply 60 %, owing to a scarcity of new rain-fed farmlands. At contemporary rates of extraction, many of the world's most essential groundwater sources in Asia, South Asia, the Americas, North Africa and the Middle East will become exhausted in the 2030s, risking a crisis in world food security and posing a direct threat to the lives of four billion people who depend on food grown using well water. Loss of groundwater also imperils the forests and landscape vegetation which depend on it, as its removal lowers the water table beyond the reach of tree and shrub roots, causing them to die: this, in turn, affects both livestock grazing and forest food production. Flying gravity instruments in satellites high above the Earth, scientists at NASA obtained a bird's eye glimpse of groundwater depletion around the planet (Fig. 7.1), finding that 13 out of 21 key basins are now stressed. They say:

> Most of the major aquifers in the world's arid and semi-arid zones—the parts of the world that rely most heavily on groundwater—are experiencing rapid rates of depletion because of water use by farms... this includes include the North China Plain, Australia's Canning Basin, the Northwest Sahara Aquifer System, the Guarani Aquifer in South America, the High Plains and Central Valley aquifers of the United States, and the aquifers beneath northwestern India and the Middle East (Voiland 2014)

Farmers all around the world are engaged in a fight for their livelihoods—and for *our* food supply—with vast cities and giant energy corporations which covet their dwindling water supplies. Swollen megacities in Asia, North America and Europe are absorbing vast quantities of new water every year. Large areas of groundwater are being drained, disrupted and polluted by oil

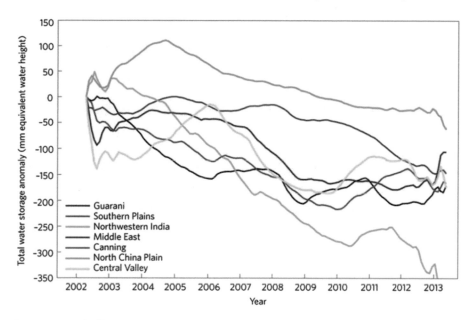

Fig. 7.1 Dwindling groundwater resources in 7 of the world's key foodbowls. Source: NASA, 2014

and gas exploration ("fracking")[1] tar-sands oil production and the excavation (and drainage) of open-cut coal mines. When fossil fuel corporations, cities and farming communities come into collision over water rights, the farmers (and hence, we consumers) almost invariably lose (See for example Finley 2012). As one Australian farmers' protest group explains:

> Our best food-producing lands and our finest natural areas are at risk from inappropriate coal and gas mining. Coal and gas exploration licences and applications cover more than 50 % of Australia and there are plans to double our coal exports and become the biggest gas exporter in the world. On the driest continent on earth, water is our most precious resource. Despite this, mining…is putting at risk our drinking water catchments, our underground water resources, and our rivers and wetlands (Impacts 2015)

Meanwhile rainfall in the world's great grain bowls is becoming less reliable as the climate changes and warms, while snowpack on high mountain chains—such as the Himalaya, Hindu Kush, Urals, Andes and Rockies,

---

[1] See for example http://earthjustice.org/features/campaigns/fracking-across-the-united-states?gclid=Cjw KEAiAgfymBRCEhpTR8NXpx1USJAAV0dQymPIy5uWZC_mNL055CpiD1F7LyEhnt4KIKJBVT-naEpRoCNiPw_wcB

which supply many of the world's great rivers and groundwater systems on which food production depends—is dwindling. (World Glacier Monitoring 2015; Brown 2011) In many countries farmers rely on these big rivers running year-round to water their crops, but if the frozen 'water tower' of mountain glaciers disappears, the rivers will only run seasonally. In some regions, such as northern India, Pakistan and Central Asia, this could halve irrigated food production.

All of these factors are placing increased stress on global food supplies at the precise moment in history when they need to increase. They are the result of poor or corrupt water allocation decisions by governments, ignorance of the resource, lack of foresight and the future-blind application of the market, setting prices on water which farmers cannot afford but others can.

Soil

The farmers of Iowa, USA, are among the world's most proficient at what they do, which is chiefly grow crops of corn and soybeans and raise livestock. But according to Iowa State University they still lose about 30 million tonnes of topsoil a year, mainly to storm erosion—and that in turn entails the loss of around US$1 billion in grain production. In a severe rainstorm, the losses can run as high as 64 tonnes of topsoil per acre. It's an unarguable fact that agriculture involves disturbing the topsoil, either by tillage or grazing, and no conventional farming system avoids it completely. However, if *good* farmers are losing so much soil, you can imagine what the not-so-good manage to lose. (Eller 2014; Cox et al. 2012) A century ago this issue hardly mattered: there was abundant new forest land to be cleared to replace any that became degraded. Now, with the world's forests and farm lands both shrinking, that is no longer the case.

In total, soil scientists estimate, the world is losing around 75 billion tonnes of topsoil a year, mainly due to food production (Wilkinson and McElroy 2006). In a separate measure of the same phenomenon, satellites have revealed the world's farm land area is shrinking at the alarming rate of about 1 % a year (Bai et al. 2008) as some of it is turned into desert, and some is buried beneath sprawling cities. If this trend continues, they warn, the world could run out of good arable land within 50–70 years (Marler and Wallin 2006; Crawford 2012).

The fault lies not with the farmers, who struggle to make a living from the declining prices which the global food chain pays for their produce. The problem lies with the merciless and future-blind economics of the global food system, which push farmers and their industries into unsustainable production in the quest for ever-cheaper food. The food in the supermarket is cheap—often one third or less what our grandparents paid for it (Van Trump 2015). But it has a hidden cost in lost soil, mined and polluted water, wasted nutrients, degraded landscapes, ruined farmers and rural communities, which is now damaging agriculture around the planet.

The solution to this issue is twofold: to reduce the economic stress on farmers and farming systems, everywhere so they can operate more sustainably (Cribb 2011) and to move to soilless or far more intensive systems of food production such as hydroponics, aquaponics, biocultures and vertical farms in urban areas (this chapter).

Terrible Waste

Ours is the first generation in human history to throw away half our food.

Between one third and a half of the efforts of the world's farmers, horticulturalists and agri-scientists, amounting to 1.3 billion tonnes of food a year worth over $1 trillion, are sent to landfill or else rot in the fields (Gustavsson et al. 2011b). The wastage is highest in the developed world, where it amounts to between 95 and 115 k per person, compared with losses in South Asia, SE Asia and Sub-Saharan Africa of 5–11 k a head. While 800 million people go hungry, the world trashes enough food to feed more than two billion. The waste is driven primarily by a global food chain which values food too cheaply, and pays farmers too little for it. In the century of global food insecurity, this is neither moral, nor economic nor sustainable.

However, since none of the rich countries which waste their food are prepared to transport it in edible condition to poor countries which lack sustenance (i.e. redistribute food) the only answer is to supply poor countries with the knowledge to produce enough of their own food. That in turn may entail their going (or being driven) down the modern, unsustainable, intensive agriculture route.

Modern high-tech agriculture, and indeed most of the human population, are completely reliant on the use of fertilisers to achieve high yields of food and feed crops. However the world currently uses about 200 million tonnes of mineral fertilisers a year to grow these crops (Heffer and Prud'homme 2014). It is probable that around three quarters of this enormous quantity of nutrients is lost, either on farm (where it can leach into groundwater, evaporate into the atmosphere, become locked in the soil or nourish unwanted weeds), or else in the form of post-harvest crop losses, food waste and discharged sewage. Attempts to double global food production by conventional farming methods also imply a probable doubling in global rates of fertiliser use, notably in Asia and Africa where soils are seriously impoverished from long cultivation. At the same time many of the world's high grade phosphate and potash reserves are running down, with those of phosphorus—an element indispensable to all life on Earth—expected to become depleted within about 50 years (Gilbert 2009; Pearce 2011). The problem is compounded by the fact that most remaining phosphate reserves are hard rock, which requires vastly more fossil energy to mine. While nitrogen for nitrogenous fertilisers is abundant in the atmosphere, producing them depends on natural gas which is a greenhouse gas and will also become scarce by mid-century. All this will drive up the price of synthetic fertilisers to unaffordable levels for most farmers, especially in the developing world. This in turn spells reduced farm production at a time of increased demand—and consequently shortages and much higher food prices for consumers.

Heavy use of nitrogen-phosphorus-potash (NPK) fertilisers over the past century has had another pernicious side-effect, the depletion of the world's best arable soils of the micro-nutrients vital to health and life. When you over-boost crop plant growth with these fertilisers, the plants in turn 'mine' the soil of these minor but essential elements, which are not being replaced. For example, a US study indicates we now have to eat five tomatoes or cauliflowers to get the same essential minerals as our grandparents, a hundred years ago, gained from eating just one (Marler and Wallin 2006). Scientists also suspect that this decline in 'nutrient density' in the modern diet may be a factor in the rising global incidence of chronic diet-related diseases.

All of this creates the potential for a nutrient crisis in the mid-century if the current approach of mining the Earth's soils and minerals continues, since all mines eventually run out. The solution is for the world to urgently recycle *all* our waste nutrients—crop wastes, food waste, postharvest losses, sewage, forestry and garden clippings and so on, back into food production. However, this will require every city and town to outlaw the disposal of food and other organic waste in landfills, to mandate its recycling into fertilisers, soil improvers, composts and other products, and to create local processing centres to do this. It will also require the development of urban agriculture on a global scale, to take advantage of the vast quantities of nutrients concentrated in the big cities.

Energy Risks

Oil is the lifeblood of modern mechanised farming. Just to feed ourselves means we each 'eat' the diesel fuel component of 66 barrels of oil a year—or about 1.3 l of diesel for every meal served. Oil is required to plant, irrigate, harvest, store, transport, process and deliver our food—typically half the oil is used on farm and half beyond the farm gate. Overall, food production accounts for around 30 % of the world's total energy use (UN FAO 2014a).

Put simply, for most of humanity: no oil, no food.

Future oil shocks thus represent one of the gravest and most immediate risks to the global food supply, especially in developed countries and megacities. These may arise as a result of gradual depletion of the world's easily-accessible fossil oil reserves—known as Peak Oil—or as a result of wars and governance failures in key producing regions, or local disasters. Peak Oil occurs when an oilfield passes the high point in the extraction of its resource and production begins to taper off: this has happened many times for individual oil fields around the world and is now happening in major oil provinces such as Saudi Arabia (Patterson 2014). Recent discoveries, such as in the Arctic, from unconventional sources like tar sands and shale oil, and from deepwater drilling have led some experts to conclude "The world is not running out of oil itself, but rather its ability to produce high-quality cheap and economically extractable oil on demand" (Kuhlman 2015). However, many of the alternatives to petroleum, such as tar sands or crop ethanol are far less energy efficient and may not be able to sustain modern industrial civilisation.

A major issue, and one which has largely escaped the attention of governments and experts, is that world production of new cars in the early 2010s grew several times faster than the volume of new oil being discovered, from all sources, conventional and unconventional (International Organisation of Motor Vehicle Manufacturers (OICA) 2013; US Energy Information Administration 2013). Such excessive growth in potential demand for oil relative to supply, if maintained, creates a risk of future oil shocks. With many of the world's farmers, especially in Asia and Africa, still in transition from manual to mechanical farming systems, both they as well as farmers in traditional western farming systems are highly vulnerable to oil scarcity or high prices. Globally, the corporatized food chain with its just-in-time approach to food distribution, is totally oil-dependent and highly susceptible to major fluctuations in oil price or supply. Governments, as a rule, have done little to insure against this, meaning that in the twenty-first century a global oil crisis may quickly explode into a global food crisis.

A fundamental paradox of energy-intensive agriculture is that the more oil it burns and land it clears, the more greenhouse gases it emits—and the less reliable this makes the very climate on which food growing depends. The present oil-based food system is thus sowing the seeds of its own potential destruction. If humanity is to avoid famine in the mid-century, it has to change.

The solution to this problem is to rapidly wean the food industry worldwide from its addiction to fossil fuels. One way to do this is to develop renewable liquid fuels, but in forms which do not compete with agriculture for land, water, energy or fertiliser. This promising alternative is explored below.

Urban Hunger Traps?

By 2050 the world's cities will be home to more than 7 billion inhabitants (WHO 2014) and cover an area of the world's best farming soils equivalent to the size of China. These gigantic cities have one terrible flaw: they cannot feed themselves. They rely on a river of trucks, planes or ships coming daily to restock the shops and markets. Much of their food travels from thousands of kilometres away, sometimes the other side of the planet. Any break in the flow—an oil crisis, a war, even a major flood or storm—and a megacity would starve within days. That food supplies might fail is something that has escaped most modern urban planners. Recent experience of events such as cyclone Haiyan in the Philippines, Hurricane Sandy in the US and the Bangkok floods of 2011 indicates that panic-buying by the local populace can strip shops of all food within 24–48 h. While most cities have emergency measures for natural disasters such as flood, fire, storm, earthquake or disease outbreak, few are equipped to survive a food emergency and most would depend entirely on outside aid.

Food, water and energy crises now constitute a major interwoven threat to the Earth's most densely-inhabited regions. While not jeopardising civilization as a whole, any megacity collapse would inevitably send shockwaves round the planet in the form of refugees, soaring food prices, shortages and economic impacts. If our current unpreparedness persists, the world is likely to witness several of these events in coming decades.

An emerging issue is the poor nutritional quality of the diet for many urban dwellers, especially those on middle and low incomes. This is the result of the industrialisation of food in the global food chain, the use of thousands of chemicals in processing and packaging it and the replacement of fresh food with highly-processed or fast food in many people's diets, heavy in salt/sugar/fat but devoid of vitamins and essential minerals and micronutrients. This has created of 'food deserts' even in relatively affluent cities and new forms of malnutrition, including obesity and diabetes in both developed and developing countries. (American Nutrition Association 2010)

The solution to all these problems is for cities to grow far more of their own food, fresh and locally, using advanced urban farming systems, discussed below, and by recycling *all* of their water and nutrients back into food production. (Cribb 2011)

Fishery Failure

The world wild fish catch peaked in 1994 and has been stagnant or declining ever since (UN FAO 2014b). Indeed, recent research indicates that the collapse in sea fisheries has been steeper, even, than the UN Food and Agriculture Organization estimates and has been shrinking at around a million tonnes per year (Pauly and Zeller 2016)—bad tidings for cats and seafood lovers alike. Despite progress in developing sustainable fisheries by a handful of countries, the take-home message is that the world is not going to double its harvest of wild protein from the oceans at the same time as food demand doubles. In fact, with 90 % of world fisheries rated as fully- or over-exploited, we will be lucky even to maintain the average catch of 80 million tonnes of wild fish. At the same time the problems of overharvesting are being compounded by the spread of ocean and coastal 'dead zones' (Chap. 3) where fish cannot survive, and by the growing flood of toxic chemicals, plastics and heavy metals (Chap. 6) which we release into the oceans and much of which ends up in the fish we eat.

Recent decades have witnessed spectacular growth in aquaculture, with the world fish and water plant harvest attaining 67 million tonnes a year by the mid-2010s. However, this highly promising industry is held back by the availability of protein and nutrient sources to feed to farmed fish. Supplies of 'trash fish' are now more often used as human food due to the scarcity of table species, while feeding a huge new global aquaculture industry on grain will only apply greater pressure to world grain supplies both for humans and for livestock like cattle, pigs and poultry. Furthermore, feeding grain to fish increases soil degradation and competes with other farm industries for energy and fertiliser. These factors present a major obstacle to the world farmed fish industry developing to anything like its true potential, which is probably in the vicinity of 200 million tonnes. If this 'feed barrier' can be overcome then farmed fish can easily become humanity's main protein staple by mid-century, surpassing all other meats and poultry.

The solution to this problem lies in the oceans themselves, as we discuss below.

Knowledge Drought

Having given birth to the immensely successful Green Revolution which doubled world food production, the world scientific effort in food and agriculture has been quietly stagnating ever since (Alston et al. 2009). Of the $1+ trillion invested globally in scientific research and development today, it is estimated that less than 5% is devoted to improving agriculture or food production. Since the human population has doubled since the 1970s while food research has declined in real terms, this means we have more than halved the brainpower which humanity puts into securing and improving the food supply. This scientific stagnation has resulted in a worldwide plateauing, in some cases actual declines, in rates of crop yield gains (Grassini et al. 2013). Food production, in short, is no longer keeping pace with demand over the long haul, which is a very dangerous trend.

The biggest declines have taken place in public sector agricultural research in the developed countries—Europe, America, Australia, even China. To some extent this has been offset by growth in the private sector. However, this development tends to favour technologies profitable to the shareholders of the technology companies, rather than technologies desired by consumers, or essential to sustainable farming, to human health or to the environment. Technologies such as genetic modification of food while promising much, have yet to deliver substantial food increases to the world's table, but in the meantime have attracted public investment away from critical areas such as soil science, soil microbiology, agronomy, entomology and traditional plant breeding. Furthermore, the growing rejection of GMO food by many consumers in Europe, Asia, Australasia and the Americas suggests that over-backing of a single technology may prove a strategic error and misinvestment, in the context of a weakening world food science effort.

The solution is for the world to redouble public investment in agricultural and food science and try to make up lost ground. Since there is strong evidence that well-fed countries suffer fewer wars whereas hungry ones suffer more wars (See for example De Soya and Gleditsch 1999; Carter 1999b), the best way to fund this would be to cut world arms budgets by 10% and invest the savings in food science (Cribb 2011). This will yield a double 'peace dividend'—by reducing the amount of weaponry in circulation and by reducing the likelihood of wars in regions afflicted by food, land and water scarcity.

Killer Diets

The modern diet is neither safe nor healthy: medical scientists estimate that today two out of every three people in the world die from a diet-related disease (WHO 2014). In affluent societies over three quarters of the population now 'die by their own hand'—the one holding the fork or chopsticks—and the lion's share of their citizens' taxation is now spent on often-unsuccessful attempts by the healthcare system to cure the incurable with drugs etc.

Put simply: the world diet has to change—to one that is fresh, diverse, healthy and which *prevents disease* instead of causing it. Driven by consumers ardent for better health, by farmers seeking improved incomes from high-quality fresh produce and by a growing army of healthcare professionals who have seen what is happening before their very eyes in hospitals and hospices across the world, food is heading for a new revolution—as a potential life-saver for billions.

The factors killing the world's people are the same as those killing its soils and waters: over-industrialisation of the global food chain and the very low prices it pays to farmers. The University of Sydney's John Crawford explains:

> The connection with health is significant. Cheap food tends to be low in protein and high in carbohydrate, which is exactly the wrong balance for a healthy society. By reducing food to a mere commodity, we have created a system that is degrading the global capacity to continue to produce food, and is fuelling a global epidemic of diabetes and related chronic disease. Obesity in the US cost 150 billion dollars – 20% of the health budget – in 2008, the latest figures available, and this huge cost will rise as the broken food system takes its toll (Crawford 2012)

Cheap food also contains traces of biocides: in the US, for example, some 6000 different chemicals are used to grow, process, preserve, extend, flavour, decorate and package food. Their combined health impact on the consumer is unknown, but since many of these substances are known toxins, carcinogens and endocrine disruptors, medical studies reporting probable health impacts are multiplying. Cheap food is also nowadays often imported from developing countries, where regulatory standards and hygiene are low, corruption is rife, pollution widespread, farmers inexperienced in chemical use and farming systems are under the thumb of avaricious food corporates. Due to globalised supermarket chains and food firms, this unhealthy food is now on everyone's platter.

The answer is straightforward, and is up to us as consumers: eat local, eat fresh, eat sustainably—and be willing to reward our own farmers much better for their care and professionalism in delivering safe, fresh, uncontaminated food of high nutritional quality.

Or else, put simply: save at the supermarket, spend at the hospice.

Climate Shock

Outside of a nuclear war or asteroid collision, the biggest shock in store for the human population in the 21st Century will be the impact of climate change on the food supply.

The Holocene climate which gave birth to agriculture some 6000–8000 years ago is now extinct (Rahmstorf 2013). It will not return: the world has changed. Two degrees of global warming—now probably unavoidable—will make harvests unreliable in most of the world's great grain bowls. In India, for example, grain yields may fall by as much as 45 % (See for example The World Bank 2013). Numerous scientific estimates indicate that, without adaptation, 5° of warming could halve global food production—at the very time we need it to double. The staple American crops, corn and soybeans, for example, are predicted to suffer yield losses of between 63 and 82 % (Schlenker and Roberts 2009). The sea level rise which warming causes will inundate most of the world's low-lying river delta regions—themselves major foodbowls. By 2030, 54 million people will be driven from their homes each year by floods, a 150 % increase on present levels (Lehman 2015).

The impact of climate on food is already being felt. The IPCC's 5th Assessment notes: "Based on many studies covering a wide range of regions and crops, negative impacts of climate change on crop yields have been more common than positive impacts (high confidence)." What this means for future food security is serious: one international scientific study finds that every 1° increase in local temperature results in a 6% decline in wheat yields (Asseng et al. 2014). If wheat crops are not specially adapted to these higher temperatures, it means the world's bread supply could shrink by as much as a third at the very time it needs to double (Natural Resources Institute of Finland 2015). If the climate warms by 4°–5°, then crop yields will need to double to offset climate losses, then double *again* to meet human demands. The sheer magnitude of this challenge is widely ignored by agricultural policy experts, who are wedded to ancient production systems. It signifies that traditional farming is not the way we will mainly feed humanity in a hotter, resource-depleted world.

This dilemma will affect everyone on Earth, if not with actual starvation then at the very least in terms of the price, availability and nutritiousness of food. It will become a principal driver of geopolitics, warfare, migration, pandemic disease and refugee tsunamis for much of the century. As the Center for Strategic and International Studies warned, some of these wars could go nuclear (Campell et al. 2007).

The solution to this quandary is for humanity to rely far less on the traditional European grain-and-meat diet, which requires a cool, stable climate—and more on a lighter, healthier Asian-style fish-and-vegetable diet utilising novel highly-intensive and indoor systems which are resilient to climate impacts, or else unaffected by them.

This isn't simply a change of diet or a food fashion. On this transition rests the future of civilisation.

Farming Transformed

We are entering one of the most exciting and promising epochs in human history: The Age of Food. The need to feed the megacities is already sparking worldwide interest in urban agriculture and horticulture (Drescher 2005). Ideas range from high-tech glass 'skyfarms' (See for example Despommier 2013; Wang 2015b) producing vegetables, fruits, small livestock to the large-scale cultivation of fresh foods on urban roofs and walls, to a renaissance in backyard, balcony, private allotment and public food garden production, to the redemption of old factory and industrial wastelands as thriving micro-farms, as in the city of Detroit (Kolasanti et al. 2012).

Aquaponic farms, where fish and vegetables are grown together in integrated systems, are sprouting from Norway, Iceland and America, to Canada, France, New Zealand and Australia (Blue Smart Farms 2014). Among recent proposals are 4000-ha vertical farm for the heart of Rotterdam (Carter 2011), a US$30 m vegetable facility for Newark, New Jersey (Wang 2015a), a $50 m-a-year fish and vegetable development for McBride, Canada (ecoTECH 2012), and a glass agri-sphere for Linkoping, Sweden (Plantagon 2016). Smart hospitals are culturing fresh vegetables on the roof to assist patients in recovering from illness (Greene 2012). Smart restaurants are offering patrons salad greens harvested just 15 min ago. (Held 2013). In Australia, Blue Farms is producing fresh fish and fresh herbs for a major national supermarket chain using conveyor-belt aquaponics (Legg-Bagg 2014). In Norway, Miljogartnieret produces 2200 tonnes of organic tomatoes and peppers in 7 months on just 7 ha (Hegelstad 2014).

These enterprises exemplify the originality and cutting-edge of sustainable, climate-proof food production systems for the twenty-first century. Their hallmarks are pinpoint management and reuse of nutrients, recycling of waste, water and energy, low or zero use of pesticides, biological pest control, hospital-style crop hygiene, specialty crops and premium quality control. Since crops are mainly cultivated indoors, they are largely immune to adverse weather or climatic conditions and can often be grown year-round. Food yields are typically 10 times or more those of field-based agriculture.

The looming world crises in arable soils and water can be overcome in two further ways—by 'desert farms' in hot, dry countries and by 'floating farms' in regions of dense population close to the sea. Desert farms, like the ones springing up around Whyalla in southern Australia (Margolis 2012), use solar energy to turn saline groundwater or seawater into fresh water to irrigate fruits, vegetables and dairy feed in regions where sunlight is plentiful but good water is scarce. They offer a promising solution to food insecurity in regions such as the Middle East, Central Asia, western China, south central India and desert Africa. Furthermore, they eliminate the need for the neo-colonial 'land-grab' of African farmers' land by Chinese, American and Arab corporations, leaving more African food for Africans (Robertson and Pinstrup-Andersen 2010; Schiffman 2013). In Asia especially, giant floating greenhouses that distil fresh water from sea water are already being designed by far-sighted architects and horticulturalists to meet the fresh food needs of seaboard megacities like Shanghai, Tokyo, Singapore or Mumbai (Schiller 2014).

Food will re-sculpt the twenty-first century megalopolis through the philosophy of 'agritecture', the incorporation of sustainable food production into building and urban design. 'Green cities', alive with vegetation, fresh food, birds and insects, will replace the polluted, soulless concrete-and-glass urbanscapes of today, enhancing both liveability, comfort and sustainability. Their cornerstone will be the recycling of all their water and organic wastes back into sustainable food production, thus dramatically lowering the city's destructive footprint on its surrounding landscapes and on wildlife. The food of such cities will effectively be climate-proof—untroubled by the weather shocks that will hammer traditional broadacre outdoor farming. They will ensure a highly-diverse, fresh and *local* food supply that never fails. By 2050, with adequate investment and R&D, urban farming can potentially supply up to half, or more, of the world's food.

This is not only essential to the megacities: it will also bring immense relief to the stress now imposed on the world's soils, water, biodiversity, landscapes and farming communities. For the first time humanity will be able to feed itself well, in abundant diversity—without plundering the natural world.

This in turn will pave the way for the regeneration of grasslands, forests, rural landscapes and communities, the locking-up of carbon and the recovery of imperilled wildlife around the world. By contracting agriculture to the best and most reliable areas over several decades, up to half of the world's farming and grazing landscapes—an area of more than 24 million square kilometres, equal to the size of North America—can be returned to wilderness, stewarded by local farmers and indigenous peoples. Given the lack of success with current national parks and conservation efforts in slowing extinction rates, only a visionary change in Earth system management such as this has the capacity to prevent the 6th Extinction of the Earth's species, described in Chap. 2 (Carey 2014). However, this is also good news for the world's beleaguered farmers: paying a small environmental levy on food—compensating for the cost it imposes on the planet—will enable them to achieve better incomes on the balance of the world's best farm land, for premium quality, clean produce grown sustainably under natural conditions, supplemented by incomes as managers and stewards of re-wilded lands.

Blue Revolution

The second great change in The Age of Food will be a 'blue revolution' cultivating fish and water plants on a hitherto undreamed-of scale. Worldwide, aquaculture produces 67 million tonnes of fish and water plants (algae) a year (UN FAO 2014b)—but this is a fraction of its potential. It will grow three-, maybe fourfold by the mid-century. By 2050 the UN FAO anticipates, the world will demand around 550 million tonnes of animal and fish protein every year (UN FAO 2013). If we were to supply this as meat we would need to grow an extra 3 billion tonnes of grain—which would require an area of arable land equivalent to three more North Americas. Given climate change and land scarcity, this is arguably impossible. So where is the feed for all these animals and fish to come from?

The answer is from the cropping of algae—or water plants. In future huge algae farms—on land, at sea and in salt lakes—will produce food for people, feed for fish and other livetock, fuel for transport, pharmaceuticals, plastics, textiles and fine chemicals (Cribb 2013). They will replace the increasingly costly extraction of fossil oil with 'liquid solar energy', oil produced fresh each day by algae using sunshine, salt water and urban 'waste' nutrients, without any net carbon emissions. More than 30 countries are already investing in what could by mid-century become the world's largest cropping industry. Algae can be farmed in tanks or ponds on wasteland, in salt lakes, in shipping containers and on floating rafts or in huge plastic bags in the ocean, where they do not compete for space with other food crops or wilderness. They are resilient to climate change. The world has 72,000 species of water plants (Guiry 2012), many containing nutrients essential to a healthy diet: they are the original source of omega three oils, betacarotene and many vitamins and minerals. These algae typically contain 30–70 % fresh oil, the balance being carbohydrate and protein which can readily be turned into human food, stockfeed or fish food.

Algal oil is, literally, liquid solar energy. This oil can be made into anything you can make from fossil petroleum—'green' fuel, plastics, textiles, chemicals, drugs, food additives. Furthermore, researchers have calculated, algae could supply the world's entire transport fuel requirement from an area of 57 million hectares—which is a bit smaller than Switzerland—and can mostly be in the ocean in any case. Even if cities use electric vehicles, oil will still be needed for farms, long-haul road, sea and air transport and other important activities. Algal oil will ensure our food system never runs out of the liquid energy it needs to grow, harvest and transport the world's food supply. And importantly, climate-friendly algal oil will help to slow and eventually reverse global warming.

Novel Foods

According to Australian agronomist Dr Bruce French there are at least 27,600 edible plant species in our world (French 2016). Since modern humans consume only a few hundred different plants, this means we have barely begun to explore our own planet in terms of what is good, healthy or delicious to eat.

Many edible plants which don't feature in the modern diet are eaten by indigenous people—but this knowledge is local, very fragile and will soon be lost. Most of these plants are vegetables—and vegetables can be produced far more quickly, using less soil, water, energy, carbon and fertiliser than other foods. They also fit the profile of a much healthier diet, which prevents disease.

Whether they are gathered by wild harvest, grown by traditional horticultural methods, raised with fish aquaponically or cultivated intensively in urban skyfarms, novel vegetables, fruits and grains in boundless diversity will form the mainstay of the future global diet. They will create new industries and jobs. They will help to employ the one billion small famers now being driven off their land by giant supermarkets, land-grabbers and global food chains (Cribb 2012).[2]

In contemplating the future of food, there are other essential trends likely to transform the diet. In 2011 Maastricht University produced the world's first synthetic sausage, and in 2013 the world's synthetic hamburger, grown from animal stem cells (Maastricht 2013). Cultured meat requires *far* less soil, water, fertiliser, energy and carbon to produce than traditional meat. It is potentially healthier and safer to eat, as well as cheaper to the consumer. It is real meat. It just never went moo.

Within a decade cultured meat could occupy the lower end of the meat market—as filler for pies, sausages, in burgers, snackfoods and the like. It is likely to appeal to the health-conscious, to 'ethical eaters' who wish to avoid exploiting animals, and those who wish to reduce their personal impact on the planet—but still enjoy a meat meal. If you doubt it will catch on, look at your clothes. Sixty years ago, nobody wore synthetic clothing—today everyone does.

Biocultures, too, will have a huge transformative effect on food in the twenty-first Century. Cell culture has long been used in agricultural and medical science but in the near future this technology is poised to emerge at industrial scale for growing healthy food. (Dal Toso and Melandri 2011) Cells from plants, fungi, microbes and other organisms can be cultured *en masse* in bioreactors—and turned into edible, nutritious and even delicious foods. Again, such production methods achieve dramatic savings in soil, water, nutrients, and energy compared with agricultural systems. More importantly, they can be specially tailored to the dietary needs of the individual consumer, based on their personal genetic risk profile, to protect us against heart disease, diabetes, cancer, or obesity. The age of the personally-specified, delicious diet is at hand.

As Asian tastes and the need to recycle nutrients catch on globally, unfamiliar foods such as insects, jellyfish and reptiles may become feature more prominently in world cuisine. Entomoculture—the farming of insects—especially has a role to play in nutrient recycling, for example by using farmed crickets to consume vegetable waste then feeding the crickets to fish or chickens for human consumption, or alternatively as a human food popular in SE Asia.

---

[2] Based on the observation that the transition from smallholder farming to large-scale, intensive agribusiness in western countries has typically resulted in <80 % of farmers leaving their farms.

The advanced technology of 3D printing also promises to play a role in novel food design. Loaded with ingredients or even basic nutrients the 'Food Printer' will combine them in the right proportions and 'print out' the chosen recipe as a ready-made dish, to order. Numerous companies around the world are working on different technologies to make food printing an industry of the future[3]. Food printing has applications ranging from on-farm, to the manufacture of precisely-specified, uniform foods, all the way through to the creation of elaborately sculptured desserts by world leading chefs.

Just as the 70s' were the age of music, the 90s' the age of the internet, we are now entering the Age of Food. Never has world cuisine been so spectacularly diverse—or so far short of its true potential. Never have the opportunities been so splendid or so great. Food is one of the most creative acts which we humans perform.

How we approach food in the will, by 2050 define the future and fate of our global civilisation.

What We Must Do

1. Develop sustainable urban food production using recycled water and organic waste in all of the world's cities
   Pathway: climate instability will drive much of farming 'indoors' in any case, while water and fertiliser shortages will also encourage the trend to 'sustainable intensification'. However urban planning (to recycle both water and nutrient waste back into food production), investment incentives (by governments anxious to avoid climate-caused food crises) and more research and development of urban farming systems and biocultures are essential.
2. Raise the next generation of humans to value and respect food far more than today.
   Pathway: this can be achieved by introducing a Year of Food in every subject in every junior school on the planet, and encouraging the food industry to help educate consumers by publicising sustainably-produced foods.
3. Through a tax on food, pay farmers and indigenous peoples to restore and maintain the world's landscapes, waters and wildlife. Restructure the economics of the global food chain to encourage sustainable production by educating consumers.
   Pathway: most governments have consumption taxes already. In this case it is proposed to apply this to food (reflecting the environmental cost of producing it) and then spend the revenue raised on paying farmers and indigenous peoples

---

[3] (see for example http://3dprintingindustry.com/food/).

to repair the damage and re-wild half the planet. For poor citizens, subsidies, food stamps or tax exemption will avoid regressivity.

4. Replace the current waste of <40 % of world food with a system that recycles all nutrients back into food production.

    Pathway: this is largely down to urban planners, who must establish in every city, 'green waste' recycling systems and prohibit the dumping of food or green waste in landfills.

5. Reshape the world diet from one that degrades the planet and our personal health to one which protects and preserves both.

    Pathway: This is achievable through better consumer health education via schools, healthcare services, the food industry (see 2, above), the media and social media, farmers' markets etc. To transform food production it is necessary for consumers to send the right economic signals to producers and farmers. It is also necessary for the medical professions to embrace preventative healthcare and nutrition to receive greater priority than chemical therapies.

6. Reinvest a tenth of world defence spending in world peace through food, particularly in agricultural, horticultural and dietary health research and development.

    Pathway: there is a need for more academic research into the issue of 'peace through food' to demonstrate to governments that the chances of war are reduced when food supplies are secure, and that investment in sustainable food pays a peace dividend. The military needs to be engaged in this discourse.

7. Introduce a 'Year of Food', teaching respect, awareness and appreciation of food, in every junior school on the planet. Empower children to educate their parents about sustainable, healthy food and caring for the ecosystem that supplies it.

    Pathway: the information to do this exists now. Numerous programs are already reintroducing the growing and preparing of simple foods to urban schools. It can be done without altering the curriculum, but merely by incorporating a food theme into every subject taught—from science and maths, to languages, to geography and social studies to sport.

8. Ensure the availability of family planning and education and healthcare for women in all societies

    Pathway: this is already well under way, but needs to go ahead faster and be coupled with equal opportunity for women and the acceptance of women as leaders in all walks of life. This presumes that women will naturally and voluntarily reduce their rate of fertility and hence, the human population, if granted the education, healthcare, power and family planning support to do so.

9. Establish a universal internet-based system for sharing food and agricultural knowledge with all farmers and consumers.

    Pathway: various organisations, public and private, are already working on this, but they need to collaborate better and more urgently.

What You Can Do

- Use the power of your consumer dollar to tell food retailers and producers that you highly value safe, healthy, sustainable, fresh food.
- Be willing to reward farmers fairly for looking after the landscape as well as producing safe, healthy, delicious produce.
- Eat fresh, eat local, eat sustainably, eat healthily. Learn what these things entail and how to distinguish between evidence-based facts and claims driven by money, unfounded beliefs or vested interests.
- Consciously avoid highly processed 'industrial food' and fast food that is unhealthy, contains too many chemicals and additives, is heavily packaged and which disrespects farmers by paying them badly. Eat for lifelong good health.
- Support family planning, education for women and other voluntary means of population reduction
- Teach your children to respect and value food, as every previous human generation until ours has done.
- Understand that our attitudes and values for food will define the human future, for good or ill.
- Grow more of your own fresh food, prepare and take pleasure in it.

8

The Urbanite (*Homo urbanus*)

> *On this Earth there are pestilences and there are victims, and it's up to us, so far as possible, not to join forces with the pestilences*
> —Albert Camus, *The Plague.*

Tim Henderson had one of the most disgusting jobs in the world to perform: to remove an enormous, stinking 40-metre long blob of fat, grease and domestic waste which had silently tumefied in one of London's main sewers beneath the upscale suburb of Chelsea. A blob so bloated it had burst through the ageing brickwork of the sewer walls. And Tim was among the lucky 'flushers', or trunk sewer technicians, who drew the chore of removing it, chunk by festering chunk.

"We see blockages all the time in household sewer pipes… but to have this much damage on a sewer almost a meter in diameter is mind-boggling," Thames Water repair and maintenance supervisor Stephen Hunt told Britain's *The Guardian* newspaper (Kaplan 2015; Ratcliffe 2015).

The blobs—known as 'fatbergs', which can weigh 15 tonnes or more and extend for 80 m—are becoming a common but largely invisible symptom of the modern metropolis, whose citizens and restaurants unthinkingly tip megatonnes of kitchen and other waste down the drain. "Fat goes down the drain easily enough, but when it hits the cold sewers, it hardens into disgusting 'fatbergs' that block pipes," Rob Smith, Thames Water's chief sewer flusher and member of the local "fatberg hit squad", told media. In 2013 New York City paid nearly $5m to purge fat buildups from its sewer network (Kaplan 2015).

In a way the 'fatbergs' are a distasteful allegory for the substances that also choke the sclerotic arteries of more than a billion urbanites around the world

© Springer International Publishing Switzerland 2017
J. Cribb, *Surviving the 21st Century*, DOI 10.1007/978-3-319-41270-2_8

and who will mostly die of the resulting heart disease or stroke. Fatbergs consist of the same unhealthy materials that predominate in the modern diet. They are the loathsome spawn of a culture of waste—of water, nutrients, food and energy—on a mega-industrial scale. They are another form of pollution, growing unseen in the darkness beneath our very feet: in London alone, such blockages flooded 18,000 homes with sewage in recent years. Their mind-boggling size speaks of the burgeoning of giant cities and waistlines as humanity abandons the countryside and its thrifty ways in favour of prodigal urban lifestyles. In themselves, fatbergs are hardly species-threatening—just they are another ugly symptom of a system going sadly wrong. They are another wakeup call.

A Hand-Made World

By the mid-twenty-first century the world's cities will be home to approaching eight billion inhabitants and will carpet an area of the planet's surface the size of China. Several megacities will have 20, 30, and even 40 million people. The largest city on Earth will be Guangzhou-Shenzen, which already has an estimated 120 million citizens crowded into in its greater metropolitan area (Vidal 2010).

By the 2050s these colossal conurbations will absorb 4.5 trillion tonnes of fresh water for domestic, urban and industrial purposes, and consume around 75 billion tonnes of metals, materials and resources every year. Their very existence will depend on the preservation of a precarious balance between the essential resources they need for survival and growth—and the capacity of the Earth to supply them. Furthermore, they will generate equally phenomenal volumes of waste, reaching an alpine 2.2 billion tonnes by 2025 (World Bank)—an average of six million tonnes a day—and probably doubling again by the 2050s, in line with economic demand for material goods and food. In the words of the Global Footprint Network "The global effort for sustainability will be won, or lost, in the world's cities" (Global Footprint Network 2015).

As we have seen in the case of food (Chap. 7), these giant cities exist on a razor's edge, at risk of resource crises for which none of them are fully-prepared. They are potential targets for weapons of mass destruction (Chap. 4). They are humicribs for emerging pandemic diseases, breeding grounds for crime and hatcheries for unregulated advances in biotechnology, nanoscience, chemistry and artificial intelligence.

Beyond all this, however, they are also the places where human minds are joining at lightspeed to share knowledge, wisdom and craft solutions to the multiple challenges we face.

For good or ill, in cities is the future of civilisation written. They cradle both our hopes and fears.

Urban Perils

The Brazilian metropolis of Sao Paulo is a harbinger of the challenges which lie ahead for *Homo urbanus,* Urban Human. In a land which the New York Times once dubbed "the Saudi Arabia of water" because its rivers and lakes held an eighth of all the fresh water on the planet, Brazil's largest and wealthiest city and its 20 million inhabitants were almost brought to their knees by a one-in-a-hundred-year drought (Romero 2015). It wasn't simply a drought, however, but rather a complex interplay of factors driven by human overexploitation of the surrounding landscape, pollution of the planetary atmosphere and biosphere, corruption of officialdom, mismanagement and governance failure. In other words, the sort of mess that potentially confronts most of the world's megacities.

In the case of Sao Paulo, climate change was implicated by scientists in making a bad drought worse. This was compounded by overclearing in the Amazon basin, which is thought to have reduced local hydrological cycling so that less water was respired by forests and less rain then fell locally. This reduced infiltration into the landscape and inflow to river systems which land-clearing had engorged with sediment and nutrients. Rivers running through the city were rendered undrinkable from the industrial pollutants and waste dumped in them. The Sao Paulo water network leaked badly, was subject to corruption, mismanagement and pilfering bordering on pillage. Government plans to build more dams arrived 20 years too late. "Only a deluge can save São Paulo," Vicente Andreu, the chief of Brazil's National Water Agency (ANA) told The Economist magazine (The Economist 2014). Depopulation, voluntary or forced, loomed as a stark option, officials admitted. Although the drought eased in 2016, water scarcity remained a shadow over the region's future.

Sao Paulo is far from alone: many of the world's great cities face the spectre of thirst. The same El Nino event also struck the great cities of California, leading urban planners—like others all over the world—to turn to desalination of seawater, using electricity and reverse osmosis filtration (Talbot 2014). This kneejerk response to unanticipated water scarcity echoed the Australian experience where, following the 'Millennium Drought' desalination plants were producing 460 gigalitres of water a year in four major cities (National Water Commission 2008)—only to be mothballed a few years later when the dry eased. By the early 2010s there were more than 17,000 desalination plants in 150 countries worldwide, churning out more than 80 gigalitres (21 billion US gallons) of water per day, according to the International Desalination Association (Brown 2015). Most of these plants were powered by fossil fuels which supply the immense amount of energy needed to push saline water through a membrane filter and remove the salt. Ironically, by releasing more carbon into the atmosphere, desalination exacerbates global warming and so helps to increase the probability of fiercer and more frequent droughts. It thus defeats its own purpose by reducing natural water supplies. A similar irony applies to the city of Los Angeles which attempted to protect its dwindling water storages from evaporation by covering them with millions of plastic balls (Howard 2015)—thus using petrochemicals in an attempt to solve a problem originally caused by ... petrochemicals.

These examples illustrate the 'wicked' character of the complex challenges now facing the world's cities—where poorly-conceived 'solutions' may only land the metropolis, and the planet, in deeper trouble that it was before. This is a direct consequence of the pressure of demands from our swollen population outrunning the natural capacity of the Earth to supply them, and short-sighted or corrupt local politics leading to 'bandaid' solutions that don't work or cause more trouble in the long run.

Other forms of increasing urban vulnerability include: storm damage, sea level rise, flooding and fire resulting from climate change or geotectonic forces; governance failure, civic unrest and civil war exemplified in Lebanon, Iraq and Syria over the 2010s; disruption of oil supplies and consequent failure of food supplies; worsening urban health problems due to the rapid spread of pandemic diseases and industrial pollution and still ill-defined but real threats posed by the rise of machine intelligence and nanoscience (Gencer 2013). The issue was highlighted early in the present millennium by UN Secretary General Kofi Annan, who wrote:

> Communities will always face natural hazards, but today's disasters are often generated by, or at least exacerbated by, human activities... At no time in human history have so many people lived in cities clustered around seismically active areas. Destitution and demographic pressure have led more people than ever before to live in flood plains or in areas prone to landslides. Poor land-use planning; environmental management; and a lack of regulatory mechanisms both increase the risk and exacerbate the effects of disasters (Annan 2003).

These factors are a warning sign for the real possibility of megacity collapses within coming decades. With the universal spread of smart phones, the consequences will be vividly displayed in real time on news bulletins and social media. Unlike historic calamities, the whole world will have a virtual ringside seat as future urban nightmares unfold.

New Plagues

From the point of view of an infectious microbe, like the flu virus, ebola, zika, cholera or drug-resistant TB, a megacity is an orgy of gourmet and reproductive opportunities. The larger the city, the more billions of human cells it harbours, on which the bug delights to dine, or in which it can multiply. Furthermore, cities have carefully equipped themselves with the most efficient means for spreading infectious microbes: international airports, schools and kindergartens, air-conditioned offices, steamy night clubs, dating agencies, sporting facilities, hospitals, pet and pest animals, insects, not-so-clean restaurants, markets and food factories, polluted water supplies and rivers, leaky waste dumps and cemeteries. From a microbe's perspective the modern city is nirvana.

It was those ancient Roman kings, the Tarquins—a dynasty that always received a spectacularly bad press from subsequent republican historians—who laid the essential foundation for the modern city when they built the Cloaca Maxima, the world's first sewer, to move the growing city's filthy waste further down the Tiber River (Hopkins 2012). Without this simple, enclosed stream draining sources of infection to a safer distance, Rome could never have flourished. The resulting reduction in disease and especially infant death rates in one of the largest concentrations of people at the time led to population growth, economic expansion and, especially, enough surplus males to maintain the standing army on which the city's subsequent ascendancy was built. One of the world's earliest examples of public health intervention, it also laid the foundations for modern urban planning—as well as the fatbergs of the future. The Cloaca Maxima was also a classic case of another ancient human tradition which still survives today: the habit of relocating a problem from A to B and then regarding it as 'solved'. When cities were relatively small, there was plenty of spare land and ocean around the world to absorb their foul emissions, they could afford to pollute and generally get away with it. But with the emergence of the megacities and a globalised economy in the modern era this has all changed. Megacities that do not self-cleanse and re-supply their resources risk drowning in their own filth, poisoning their citizens and cultivating waves of pollution and infectious disease which can then travel internationally in a matter of hours.

The World Health Organisation identifies 14 major pandemic disease threats to the global population: avian influenza, cholera, emerging diseases (e.g. nodding disease), Hendra virus, pandemic influenza, leptospirosis, meningitis, Nipah virus, plague, Rift Valley fever, SARS, smallpox, tularaemia, haemorrhagic fevers (like the Ebola and Marburg viruses), hepatitis and yellow fever (World Health Organization 2015a). To this formidable panoply of scourges it adds the worldwide emergence of a new wave of drug-resistant organisms, such as tuberculosis, golden staph, streptococcus, salmonella and malaria, which pose a rising hazard to human health not only from the diseases they cause that resist treatment, but also from the accompanying loss of antibiotic protection for surgical procedures, cancer therapies and the like. "Epidemics are common occurrences in the world of the 21st century," WHO explains. "Every country on earth has experienced at least one epidemic since the year 2000. Some epidemics, such as the H1N1 2009, Avian Flu and SARS pandemics, have had global reach, but far more often, and with increasing regularity, epidemics strike at lesser geographic levels. Devastating diseases such as the Marburg and Ebola haemorrhagic fevers, cholera, plague, and yellow fever, for instance, have wreaked havoc on regional and local scales, with much loss of life and livelihoods" (World Health Organization 2015b). By redistributing disease-carrying mosquitoes worldwide, as in the case of the Zika virus, climate change is also augmenting the risks of pandemics, according to the *New York Times*: "Recent research suggests that under a worst-case scenario, involving continued high global emissions coupled with fast population growth, the number of people exposed to the principal mosquito could more than double, to as many as 8 billion or 9 billion by late this century from roughly 4 billion today" (Gillis 2016).

Of the 60 million or so people who die in our world each year, as many as 15 million die from an infectious disease—the rest perishing chiefly from lifestyle diseases and a much smaller number from accidents and wars (World Health Organization 2014). This underlines the dramatic change in the modern era, in which infectious disease has become a far less common cause of death than was the case throughout most of human history—thanks chiefly to the advent of vaccines, antibiotics and sound public health measures. It also highlights the dramatic rise in deaths from self-inflicted disease and the almost complete failure, so far, of preventative medicine. However, a diet of disaster movies and highly-coloured news reports has left the public with the erroneous impression that the risk from infectious disease is much greater than, for example, the risks posed by our own poor food choices, air or water pollution, whereas the opposite is in fact true. If there exists an *Andromeda*

Strain-style agent capable of wiping out the whole of humanity,[1] it has yet to come to the attention of science—and, for good biological reasons, it probably doesn't exist unless somebody artificially creates it: natural organisms seldom eliminate all their hosts, as to do so is not a good strategy for their own survival. Instead they attenuate and adapt—a lesson humans need also to contemplate.

Viewed from the perspective of a direct threat to the existence of civilisation or the human species as a whole, the risk from infectious disease *per se* comes a long way behind that of nuclear war, climate change, global toxicity, famines and some of the other technological perils described in this chapter. However, pandemics frequently arise as a synergetic consequence of war, famine, poverty, mass migration, climate change, ecological collapse and other major disasters and therefore play an amplifying role in endangering the human future. The classic case was the 1918–1919 influenza outbreak which arose in the immediate aftermath of World War I largely as a consequence of the world-wide movement of soldiers and refugees at a time when many populations were weak from hunger. The 'Spanish' flu infected an estimated 500 million people worldwide, killing between 20 and 50 million of them.[2]

The organisms which pose the greatest pandemic dangers in the twenty-first century—such as avian flu, Ebola, HIV and SARS—mostly originate in wild or domesticated animals and often arise out of some sort of environmental decline. As human numbers grow and people push into areas formerly dominated by wildlife and forests, more of these zoonotic diseases (animal-sourced infections) will probably transfer into the human population: as we replace their natural hosts with large concentrations of people, the viruses have little option other than to jump species, if they are to survive themselves. However, the very fact that their likely origins are understood, if not always precisely known, makes it possible to establish detection, early warning and prevention systems, which are the current goal of world health authorities (McCloskey et al. 2014). In the second category of threat are diseases which transfer into humans from domesticated livestock—seasonal influenza outbreaks, the Nipah, Hendra and MERS viruses and food-borne infections like *E. coli, salmonella* and *listeria*. Here too, early detection and prevention hold the key to arresting pandemics.

Unknown diseases can strike without warning out of the world's fragmenting environments, as shown by the cases of HIV, Ebola, Nipah and Zika. HIV originated with SIV, a relatively harmless virus of African monkeys and

---

[1] The *Andromeda Strain* was a classic 1971 science fiction movie. http://www.imdb.com/title/tt0066769/
[2] 1918 Flu Pandemic. History Channel. http://www.history.com/topics/1918-flu-pandemic

apes which crossed into humans, who had no resistance to it, during the mid-twentieth century—nobody yet knows how for sure (Cribb 2001)—and by the early 2010s had claimed 25 million lives and infected a further 35 million individuals, most of whom will eventually die from it. However, the development of preventive strategies, education, better drug therapies and vaccines all promise to reduce the toll. Ebola, a frightening infection in which victims leak contagious blood and convulse, is thought to originate in bats or rodents and first emerged as a human disease with outbreaks in the Congo in 1976 and 1979. A major eruption in West Africa in 2013 infected 25,000 people within a year and killed 10,000. Despite the alarming speed of its onset, the infection was contained and most patients with access to good healthcare made successful recoveries (The Economist 2015b). The experience in these cases suggests that new pandemic diseases crossing into humans from wildlife can be contained and, even if they have large initial local death tolls, they do not pose an existential threat to humanity at large. However, it is far easier to limit their impact by taking effective medical, public health and quarantine action close to the point of origin—and this depends heavily on the local government, its skills and resources, and its willingness to co-operate with others, nationally, regionally and globally.

The best candidate for a twenty-first century version of the 'Black Death' is still the flu virus, in one of its newer evolutions, its close relatives such as the avian influenza H5N1, or SARS. The reason is that these viruses can be transmitted in airborne droplets from coughs and sneezes, not just in bodily fluids as is the case with HIV and Ebola. Robert Webster, a professor of virology division at St Jude Children's research hospital in the UK explains: "Just imagine if the Ebola outbreak in West Africa was transmitted by aerosol. If flu was just as lethal. If H5N1 [avian flu] was as lethal in humans as it is in chickens – and studies have shown that it only takes about three mutations to make it highly lethal. It's not out of the realms of possibility" (Woolf 2014). One reason flu mutates so often is because the virus is constantly cycling between different poultry, pig and human populations: each host presents it with fresh genetic challenges, forcing it to evolve novel strains in order to adapt: sometimes these prove more infectious, or more deadly, making it a very clever virus. Australian virologist the late Frank Fenner—one of the heroes of the world campaign to eliminate smallpox—once stated that a neuropathogenic strain of avian flu (one which infects the brain and central nervous system of birds and kills them rapidly) was the plague he most feared because it was both highly infectious and highly lethal: in theory, a sneeze on an airliner could kill most of the passengers. So far, however, such a strain has yet to cross from birds to humans. Projections for a major outbreak of a new and deadlier strain of flu suggest that it would infect between 100 and 1000 million people and kill from 12 million to 100 million of them, depending on how quickly and effectively the outbreak was suppressed (Klotz and Sylvester 2012).

Two other major killers which could potentially account for a large part of the population if widely released are smallpox and SARS. Smallpox, one of the worst human plagues throughout history which used to kill up to two million people a year, was declared eradicated in 1980 following a global immunization campaign led by WHO. The last known natural case occurred in Somalia in 1977. Since then, the only other reported cases were caused by a laboratory accident in 1978 in Birmingham, England, which killed one person and caused a limited outbreak (World Health Organization 2015c). However, the virus has not been eliminated from the Earth: both the United States and Russia are thought to maintain stocks in their biological warfare laboratories, perpetuating the risk of either a laboratory escape or deliberate release. To the end of his life, Fenner campaigned for the complete destruction of all smallpox virus stocks worldwide.

In a far-sighted paper published in 1996 Paul Ehrlich and Gretchen Daily argued that, while a pandemic would be a horrible way to end the human population surge, reversing human population growth *voluntarily* is a wise and sensible way to reduce the risks of future pandemics (Daily and Ehrlich 1996).

Man-Made Killers

The potential for the rapid global spread of a new plague agent was highlighted in 2002–2003 with the outbreak of severe acute respiratory syndrome (SARS). In this "a woman infected in Hong Kong flew to Toronto, a city with outstanding public health capabilities. The woman caused infections in 438 people in Canada, 44 of whom died." Ultimately the disease infected 8000 people worldwide and killed nearly 800 of them. "What if the next infected person flies to a crowded city in a poor nation, where surveillance and quarantine capabilities are minimal? Or to a war zone where there may be no public health infrastructure worthy of the name?" asked Lynn Klotz and Edward Sylvester, writing in *The Bulletin of Atomic Scientists*. Checking around, they identified no fewer than 42 laboratories worldwide which keep live stocks of potential pandemic pathogens (PPPs) such as SARS and the 1918 flu virus for 'scientific and military purposes' (Klotz and Sylvester 2012).

The risks of a man-made plague were highlighted in a scientific row which broke out in 2014 over the work of Wisconsin University microbiologist Yoshihiro Kawaoka who, as part of an experiment to understand the evolution of flu viruses, had deliberately engineered a strain of the 2009 killer H1N1 virus into mutated forms to which humans were completely susceptible and had no immune protection. Professor Kawaoka claimed his mutant strains were intended purely to help in the development of vaccines, but other scientists pointed out that if they either accidentally escaped or were deliberately released from his medium-security laboratory, the effects could be horrendous (Connor 2014). The episode underlined the lack of ethical oversight of scientists worldwide engaged in designing new and potentially deadly life-forms.

That the intentional release of a pandemic agent from even the most secure government facility can happen was proved beyond doubt by the 2001 American case in which five people died and 17 became sick following the mailing of anthrax spores to offices of the US Senate and to media. When analysed, the anthrax turned out to be the 'Ames strain', a type specifically engineered by American biowarfare scientists from a microbe found in a Texan cow, and subsequently distributed to 16 laboratories across the US. After intensive investigation by the FBI, it was concluded that the microbes had been mailed by a mentally-unbalanced employee of the US biowarfare facility at Fort Detrick, Maryland, in the wake of the 9/11 terrorist attacks, to highlight America's vulnerability to this form of attack and to frighten the Congress into increasing funding for biowarfare research. He succeeded. However, since the suspect committed suicide soon afterwards, his motives were never clarified beyond doubt (US Federal Bureau of Investigation 2016). The important take-home lesson from the event is that no laboratory anywhere in the world, no matter how secure, is proof against malicious distribution of plague agents, whether natural or artificial, by a crazy or fanatical employee, a government acting in its perceived national interest, an undercover enemy agent or just by plain accident. All biowarfare laboratories- and indeed, many ordinary biotech labs – thus represent an ongoing existential threat to humanity whose safety, like that of nuclear materials, cannot *ever* be guaranteed.

This was highlighted in early 2016 when James Clapper, U.S. director of national intelligence, issued a warning that even gene editing (such as by the technology known as CRISPR) should be added to the list of weapons of mass destruction, adding that it "increases the risk of the creation of potentially harmful biological agents or products". (Regalado 2016). Other scientists warned that genetically modified lifeforms could be used to target specific groups of humans carrying certain genes, or if released in agricultural 'designer crops' might result in uncontrollable plagues. They cautioned that gene editing technology is far cheaper and easier to access than nuclear or chemical weapons.

Of the dangers of 'synthetic biology'—the artificial making of novel life-forms—the Global Challenges Foundation says "The design and construction of biological devices and systems for useful purposes, but adding human intentionality to traditional pandemic risks..." constitute one of the 12 major existential threats it identifies to humanity in its 2015 report:

> Attempts at regulation or self-regulation are currently in their infancy, and may not develop as fast as research does. One of the most damaging impacts from synthetic biology would come from an engineered pathogen targeting humans or a crucial component of the ecosystem.
>
> This could emerge through military or commercial bio-warfare, bioterrorism (possibly using dual-use products developed by legitimate researchers, and currently unprotected by international legal regimes), or dangerous pathogens leaked from a lab. Of relevance is whether synthetic biology products become integrated into the global economy or biosphere. This could lead to additional vulnerabilities (a benign but widespread synthetic biology product could be specifically targeted as an entry point through which to cause damage) (Global Challenges Foundation 2015).

Machine Minds

In 2014 the world received a startling wakeup call when eminent British cosmologist Stephen Hawking, one of the world's best-known scientists and a man who has personally benefitted from super-smart technologies to overcome the physical handicaps imposed by his motor neurone disease, uttered a warning that artificial or machine intelligence could be the undoing of humanity. "The development of full artificial intelligence could spell the end of the human race," he told the BBC. "It would take off on its own, and redesign itself at an ever increasing rate," he said. "Humans, who are limited by slow biological evolution, couldn't compete, and would be superseded" (Cellan-Jones 2014).

It wasn't a new thought: science fiction writers have been grappling with the potential for conflict between human and machine intelligence for decades: it was a key theme of Isaac Asimov's robot stories written between the 1940s and 1960s; it was central motif in Stanley Kubrick's 1968 epic film *2001: A Space Odyssey*, in which HAL, the suavely paranoiac computer, tries to eliminate the human crew of a starship after concluding they are a threat to his mission. But in the words of Hawking, who has used latest generation AI to enhance his thought, word and speech contact with fellow humans and was impressed by its ability to interpret his wishes, it held a certain arresting quality.

Hawking wasn't alone. Tesla Motors and SpaceX CEO Elon Musk, regarded as one of the world's technological visionaries, also expressed deep disquiet.

Commenting on the emerging power of internet-based artificial intelligence he told a group of science thinkers calling itself the Reality Club: "The risk of something seriously dangerous happening is in the five-year timeframe. 10 years at most. Please note that I am normally super pro technology, and have never raised this issue until recent months. This is not a case of crying wolf about something I don't understand" (Rosenfeld 2014). Elaborating on this comment in a talk at Massachusetts Institute of Technology he said: "I think we should be very careful about artificial intelligence. Our biggest existential threat is probably that … There should be some regulatory over-sight at the national and international level, just to make sure that we don't do something very foolish. With artificial intelligence we are summoning the demon. In all those stories where there's the guy with the pentagram and the holy water, it's like, he's sure he can control the demon. Didn't work out."

Like many two-edged technologies before it, AI promises to insinuate itself into our hearts, minds and wallets by taking over all the hard, dirty, inconve-nient, boring and costly tasks that humans prefer not to do—and few of us have the penetrating gaze of a Hawking or a Musk to see where it all may lead. As with all new technologies, its boosters talk it up: its intelligent critics are heard far more rarely. Of this powerful new technology, the journal Scientific American said "Like next-generation robotics, improved AI will lead to sig-nificant productivity advances as machines take over—and even perform bet-ter—certain human tasks. Substantial evidence suggests that self-driving cars will reduce the frequency of collisions and avert deaths and injuries from road transport, because machines avoid human errors, lapses in concentration and defects in sight, among other shortcomings. Intelligent machines, having faster access to a much larger store of information and the ability to respond without human emotional biases, might also perform better than medical professionals in diagnosing diseases" (Meyerson 2015).

The issue received a major airing in January 2015, when over 4000 of the world's leading technological minds—including Hawking and Musk—signed an open letter to the Future of Life Institute, which stated:

> There is now a broad consensus that AI research is progressing steadily, and that its impact on society is likely to increase. The potential benefits are huge, since everything that civilization has to offer is a product of human intelligence; we cannot predict what we might achieve when this intelligence is magnified by the tools AI may provide, but the eradication of disease and poverty are not unfathomable.
>
> Because of the great potential of AI, it is important to research how to reap its benefits while avoiding potential pitfalls. The progress in AI research makes

it timely to focus research not only on making AI more capable, but also on maximizing the societal benefit of AI. We recommend expanded research aimed at ensuring that increasingly capable AI systems are robust and beneficial: our AI systems must do what we want them to do (Russell et al. 2015).

Although this sounds a bit like asking chemists to come up with better antibiotics but not monkey with poison gas or high explosives, or asking physicists to design better electronic devices but not build better nuclear bombs, it does at least inject the issue of ethics into the early-stage development of a potentially plenipotent and disruptive new technology.

Propelling such concerns is the sharp increase in the use of by various countries of robot vehicles, primarily airborne drones, capable of dealing death to those with whom their operators disagree in opinion, interest, politics, belief or culture—as well as to disturbingly large numbers of innocent bystanders, or 'collateral damage'. This has prompted a group of international scientists and peace activists to form the *Campaign to Stop Killer Robots*, which demands a moratorium on all new 'autonomous executions' until international law has been developed to deal with the issue[3]. The campaigners explain:

Rapid advances in technology are resulting in efforts to develop fully autonomous weapons. These robotic weapons would be able to choose and fire on targets on their own, without any human intervention. This capability would pose a fundamental challenge to the protection of civilians and to compliance with international human rights and humanitarian law.

Several nations with high-tech militaries, including China, Israel, Russia, the United Kingdom, and the United States, are moving toward systems that would give greater combat autonomy to machines. If one or more chooses to deploy fully autonomous weapons, a large step beyond remote-controlled armed drones, others may feel compelled to abandon policies of restraint, leading to a robotic arms race (Campaign to Stop Killer Robots 2015).

The killer robots are a fresh case where technology has outrun society and its ability to manage and regulate it. Remote-control military drones were barely in use for a decade, and were still unfamiliar to most of the world's citizens, before technicians were hard at work developing pieces of machinery capable of roaming at will and making their own decisions, under certain rules, about whom to murder. By the mid-twenty-first century such machines will be a commonplace in the military arsenals, police forces and security agencies

---

[3] See http://www.stopkillerrobots.org/2015/03/ccwexperts2015/.

of most countries and maybe even of multi-national corporations—on the pretext of 'better security'. Like the warhorse, the musket and the aircraft in ages gone by, 'mindless' machine killers could become a tactical game-changer, capable of hunting down individuals, menacing entire nations, corporations, regions, cities, leaders, executives or systems of belief and—in the hands of a malignant group or country—of threatening global civilisation as a whole.

However, the greater risk from AI may stem less from autonomous weapons, which operate to some extent under human direction, than from machine intelligence which might seek—for reasons of its own—to dominate, supplant or eradicate humans. Although this may sound like science fiction, it is the issue that so alarmed Hawking and Musk and is based on technologies which already exist or else are now in development. The Global Challenges Foundation explains:

> The field [of AI] is often defined as "the study and design of intelligent agents", systems that perceive their environment and act to maximise their chances of success. Such extreme intelligences could not easily be controlled (either by the groups creating them, or by some international regulatory regime), and would probably act to boost their own intelligence and acquire maximal resources for almost all initial AI motivations.
>
> And if these motivations do not detail the survival and value of humanity, the intelligence will be driven to construct a world without humans. This makes extremely intelligent AIs a unique risk, in that extinction is more likely than lesser impacts. There is also the possibility of AI-enabled warfare and all the risks of the technologies that AIs would make possible. An interesting version of this scenario is the possible creation of "whole brain emulations", human brains scanned and physically represented in a machine. This would make the AIs into properly human minds… (Pamlin et al. 2015).

The Foundation points out that such risks are not standalone; very often they intersect with, compound or trigger other risks in a domino-like chain. The risks from machine intelligence, for example, could easily complement and exacerbate threats from nanotechnology and biotechnology, producing a technology-dominated environment in which mere humans could not survive: for example, the use of drones to distribute viruses engineered to attack only humans carrying a particular set of genes. Of all the various risks facing humankind in this century, the Foundation rates artificial intelligence as the most technologically difficult to overcome, and the hardest of all to form partnership to oppose it, since so many people may have vested interests in its development. In short, the control of AI is liable to prove as problematic, disputed and intractable as the control of the Earth's climate, the control of nuclear weapons or toxic chemicals.

The precise process whereby machine intelligence would eliminate humanity is not described in any of these scenarios, but the common concern is that any AI created by humans would inherit both our own competitive instincts and ruthlessness, and unlike humans this would not be moderated by a 'moral' obligation to protect our species. It may therefore be motivated to eliminate all potential competitors or perceived risks to its own survival, including its creators. The still unanswered question in all this is the Asimovian one: can a machine be endowed with morals?

Nanocracy

A second dimension in which the march of technology imperils the human future is through the rise of the 'nanocracy',[4] a condition in which close surveillance and information about individuals throughout the whole of their lives will be maintained by a network of governments, commercial corporations and law enforcement agencies (Cribb 2007).

As whistleblowers Mark Snowden, Chelsea Manning and Julian Assange exposed, modern society and all who dwell in it are already potentially subject to intensive surveillance (Pope 2014). All of our financial, computer and mobile phone records, our health details, purchasing decisions, travel, tastes, hobbies and preferences, our appearances on security cams in shops, offices, taxis and public places all over the modern city, are available to the state—and many of them to private corporations equally powerful. Testifying to the rapid spread of surveillance devices, as early as 2013 Britain alone already had six million CCTV cameras—one for every 11 citizens—according to the British Security Industry Authority (BSIA). Our smart phones, satnav vehicles and airlines can report wherever we go with them. 'Intelligent' TVs, voice-controlled household devices and smart phones can potentially monitor, record and report our conversations and utterances even in the privacy of our own homes (BBC 2015). Our computers can scan our faces and work patterns for signs of boredom, resentment or dissidence. Technologies to interpret our brain patterns are already in their infancy, as Hawking has warned. All that is missing are computers capacious, fast and powerful enough to store, retrieve and interpret every piece of data on each individual from the moment of birth to the moment of death. These are now just around the corner, thanks to quantum technology.

---

[4] The neologism is derived from nanos, Greek for a dwarf, which is the stem of the words nanotechnology and nanometre and describes particles which are ultra-small, and kratos, Greek for power. It means power conferred by very small devices and the people who operate them.

A quantum computer is a device that goes to the next level of super-miniaturisation, using quantum particles (or qubits) which can exist in several superimposed states instead of the familiar binary digits (or bits), which exist in only two. The result is a device of massively more speed, power and memory capacity than conventional technology, or colloquially, 'a supercomputer the size of a room in a matchbox'. Researchers from the University of New South Wales, who created the world's first working quantum bit in 2012 (University of NSW 2012), told media at the time the world's first quantum computer was probably only 5–10 years away. Dr Andrea Morello said that quantum computers "promise to solve complex problems that are currently impossible on even the world's largest supercomputers: these include data-intensive problems, such as cracking modern encryption codes, *searching databases* (author's italics), and modelling biological molecules and drugs." Google and NASA claim to have built the most powerful computer ever—the D-Wave 2X—trumpeted as a major breakthrough for artificial intelligence (NASA 2015). Wall Street and banks like Goldman Sachs are investing in quantum computing, in a race to turn atomic particles into fast cash (Bloomberg 2015). Airbus is using an early version to design jet aircraft of the future (The Telegraph 2015). IBM and the US intelligence research agency IARPA are building the most powerful spying machine in history (IBM 2015).

By the 2030s, thanks to quantum computing and the universal spread of the internet and electronic devices such as smart phones and closed-circuit cameras, it will probably be feasible to observe and monitor virtually every individual in society for most of their life, automatically and without their consent: our genetic details and unique identifiers like personal smell or other biometrics, all we do and say or is done or said to us, everywhere we go and everyone we meet, all our financial transactions, private documents and photos, our unique brain patterns and biological indicators, all the vision we generate, every keystroke or touch on a mobile device, every website we visit, TV program we watch or book we read. This can potentially be stored, mined and sifted at light-speed using 'quputers'—and interpreted by artificial intelligence directed according to the purposes of the person (or intelligence) who authorised the search. For those who may attempt to isolate themselves from this universal electronic espionage, drones or swarms of microscopic 'nano-bots' will provide close surveillance (Motherboard 2014).

The Orwellian notion of a single, centralised 'big brother' surveillance brain is misplaced in the modern world. In reality, the information on individuals in the developed world already exists in hundreds, even thousands of separate databases, most of them owned by the private sector—your bank, your Facebook or email account, your internet service provider, your phone company, your car firm, your supermarket, doctor, golf club or travel agent. By the 2030s these will become retrievable and searchable in microseconds by any agency or corporation with the power to do so—and a quantum computer to do it. Advanced data mining and pattern recognition technology will enable 'targets' to be picked out of the population on the basis of their words, thoughts, habits and deeds automatically, without the individual ever having previously come to the attention of law enforcement, security services, political or religious 'thought-police' or commercial marketers. And once you have been selected as a target it will be almost impossible to get off the database. The oft-repeated claim that 'the innocent have nothing to fear' is nonsense: everyone, guilty or innocent, will potentially be subject to unblinking, 24/7 AI scrutiny throughout their lives.

These are, of course, no less than the enabling technologies for a global surveillance state—though nobody is admitting as much. While it is logical that a complex society of ten billion people requires more laws, regulations and enforcement that a nineteenth century world of half a billion humans, the advent of quantum surveillance will over-ride and eliminate most aspects of individual freedom. Without strict safeguards, transparency and public oversight, it could potentially render everyone, in effect, state property. On present trends, this will probably be accomplished with the co-operation of the private sector, via internet companies and banks, and with the gullible consent of voters reassured by government claims that spying on everyone is 'essential to national security'. With many transnational corporations now larger, wealthier and more powerful than individual countries or governments, one of the chief and most intrusive objectives of universal surveillance will be marketing—to precisely target every individual with an avalanche of products and services to anticipate their every whim, before they even know they have it. And finally, political parties and religious bodies may exploit the technology not only to spy on their opponents but to ensure the loyalty of supporters, who may then be coerced by threats to expose aspects of the private lives. This is the dawn of the nanocracy, the rule of the Dwarf Lords (see Pamlin et al. 2015).

Like all advanced technologies—and despite all the self-serving hype by the scientists working on it—there is no guarantee such omnipotence will be used wisely, benignly, ethically or well, be regulated, publicly supervised or even its details widely known. Indeed, the odds are it will first be employed by political, economic and religious elites to spy on and control those they deem a threat to their power, beliefs, wealth or freedom of action—or else an opportunity to spot customers, recruits or agents of influence. Edward Snowden, who witnessed the birth of the secretive age of universal espionage and blew the whistle on it, told Australia's ABC in May 2015 that the power to search both our content and metadata is "incredibly empowering for governments, incredibly disempowering for civil society". It could lead to what he termed a 'turnkey tyranny' in which governments claim to follow due process but secretly ratchet-up their level of intrusion into private lives without disclosing it. "They are collecting information about everyone, in every place, regardless of whether they have done anything wrong," he warned (Snowden 2015).

While most people will regard such electronic intrusion mainly as threats to individual liberty or privacy, there is in fact a far more dangerous aspect to them, which affects the fate of our species. One of the most striking lessons from communism, Nazism, McCarthyism, Jacobinism or the religious fanaticism of the past two centuries is the way they enforced surveillance on their societies, compelling citizens to inform on one another, and driving individuals to self-censor even to the point of suppressing private thoughts contrary to the prevailing doctrine.

The risk such a development on a universal scale poses to the human future in the twenty-first century is its potential to chill or prevent the very debate and change which are vital to our survival. Evidence that surveillance can discourage public discussion or the expression of opinion has already appeared in a study by Wayne State University's Elizabeth Stoycheff which found "the ability to surreptitiously monitor the online activities of … citizens may make online opinion climates especially chilly", adding "While proponents of (mass surveillance) programs argue surveillance is essential for maintaining national security, more vetting and transparency is needed as this study shows it can contribute to the silencing of minority views that provide the bedrock of democratic discourse" (Stoycheff 2016).

Many people are by nature explorers of new ideas, adventurers, challengers of accepted opinion, reformers, liberals, researchers, conservationists, pioneers, creators and innovators. These gifted individuals have led every major social and technological transformation since civilization began. They are the foil to our natural conservatism and apathy, the navigators and sources of inspiration in the human ascendancy. Progressive, prosperous and dynamic societies rely on such individuals to inspire and lead us to greater, bolder, wiser futures.

However, under the nanocracy such people will be easily picked out and 'discouraged', especially if the changes they propose threaten those who most profit from the *status quo*. Even if they are not directly censored, most people will self-censor rather than invite scrutiny. Historically, reformers, visionaries and dissidents from Socrates and Jesus to Galileo, Martin Luther King and Nelson Mandela often pay a high personal price. Under the nanocracy such people won't even have the opportunity. They will be quietly identified by AI and hushed long before they have a chance to cause trouble.

A human race deprived of its radicals, visionaries, liberals, evangelists, innovators and adventurers will be a lobotomized species, more like a termite mound than a society. It may be stable, organised and industrious—but it will also be less progressive, less creative and less resilient, because it would tend to suppress warning voices and views that contest social norms or which argue for reform. It will be a species less able to avoid the main existential threats because—as with climate change and pandemic poisoning—to do so may threaten the self-interest of ruling elites.

The advent of quantum computers and universal surveillance may thus herald a profound fork in the path of human evolution, creating a species less wise, less fit for survival at the precise moment in history when that survival is most in play (Cribb 2016).

The Wealth Divide

Worldwide, while there is abundant evidence that humanity is becoming wealthier and achieving higher living standards as a whole, there is also evidence that wealth is being distributed less evenly across many societies and is concentrating in fewer hands: to quote the old saw, the richer are getting richer and the poor—relatively—poorer. The World Bank maintains an index which ranks countries according to their income equality/inequality (World Bank 2015b) which tends to bear this out, while the international aid agency Oxfam argues that half the world's wealth is now held by just 1 % of its people.

> These wealthy individuals have generated and sustained their vast riches through their interests and activities in a few important economic sectors, including finance and insurance, and pharmaceuticals and healthcare. Companies from these sectors spend millions of dollars every year on lobbying to create a policy environment that protects and enhances their interests further. The most prolific lobbying activities…. are on budget and tax issues; public resources that should be directed to benefit the whole population, rather than reflect the interests of powerful lobbyists (Hardoon 2015).

According to The UK *Guardian*, in 2014, 80 individuals on Earth controlled more wealth than the poorest 3,600,000,000: (Elliott 2015). The *Credit Suisse Wealth Report* in 2015 came up with a similar estimate, that 1 % of the population controlled half the household assets in the world (Credit Suisse Research Institute 2015). In his book *Capital in the 21st Century*, economist Thomas Piketty showed that income inequality in North America, Britain and Australasia had climbed steadily for three decades, and by the 2010s was back on a par with where it was in the 1920s–1930s! (Piketty 2014). In the United States, the top 1% of earners controlled almost one dollar in every five of the nation's income (up from 8% in 1980 to nearly 18% by 2010). The United Kingdom's rich share rose from 6 to 15%, while Canada's grew from 8 to 12%. Many commentators have been quick to attribute the rise of extremist politics and demagogic figures to the disillusion among voters over their dwindling share of national prosperity—since, as the New York Times put it: "the wealthy bring their wealth to bear on the political process to maintain their privilege" (Porter 2014).

The argument that income inequality leads to legislative stalemate and government indecision was advanced by Mian and colleagues in a study of the political outcomes of the 2008–2009 Global Economic Recession (Mian et al. 2012), stating "…politically countries become more polarized and fractionalized following financial crises. This results in legislative stalemate, making it less likely that crises lead to meaningful macroeconomic reforms." It also affects intergenerational cohesion, explains Nobel economics laureate Joseph Stiglitz: "These three realities – social injustice on an unprecedented scale, massive inequities, and a loss of trust in elites – define our political moment, and rightly so…. But we won't be able to fix the problem if we don't recognize it. Our young do. They perceive the absence of intergenerational justice, and they are right to be angry" (Stiglitz 2016).

From the perspective of the survival of civilization and the human species, financial inequality does not represent a direct threat—indeed most societies have long managed with varying degrees of income disparity. Where it is of concern to a human race, whose numbers and demands have already exceeded the finite boundaries of its shared planet, is in the capacity of inequality to wreck social cohesion and hence, to undermine the prospects for a collaborative effort by the whole of humanity to tackle the multiple existential threats we face. Rich-against-poor is a good way to divert the argument and so de-rail climate action, disarmament, planetary clean-up or food security, for instance.

Disunity spells electoral loss in politics, rifts between commanders and their troops breed military defeat, lack of team spirit yields failure in sport, disharmony means a poor orchestra or business performance, family disagreements often lead to dysfunction and violence. These lessons are well-known and attested, from every walk of life. Yet humans persistently overlook the cost of socioeconomic disunity and grievances when it comes to dealing with our common perils as a species.

For civilisation and our species to survive and prosper sustainably in the long run, common understandings and co-operation are essential, across all the gulfs that divide us—political, ethnic, religious *and* economic. A sustainable world, and a viable human species, will not be possible unless the poverty and inequity gaps can be reduced, if not closed. This is not a matter of politics or ideology, as many may argue: it is the same lesson in collective wisdom and collaboration which those earliest humans first learned on the African savannah a million and a half years ago: together we stand, divided we fall.

It is purely an issue of co-existence and co-survival. Neither rich nor poor are advantaged by a state of civilisation in collapse. An unsustainable world will kill the affluent as surely as the deprived.

What We Must Do

1. Replan the world's cities so they recycle 100% of their water, nutrients, metals and building materials

   Pathway: primarily the role of urban planners and civic leaders, many have already begun to develop 'sustainable cities'. These cities are sharing their knowledge, technologies and experiences with one another round the planet via the internet. This is placing cities, often, far in advance of nations in dealing with issues such as climate, water, energy, recycling etc. Probably the most useful development would be a virtual 'Library of Alexandria' through which all urban plans, ideas, technologies, advice and solutions can be shared at lightspeed to cities all around the globe. Partnering between advanced and underdeveloped cities will help. The recycling of water and nutrients is top priority.

2. Stop destroying rainforests and wilderness, which forces animal viruses to take refuge in humans.

   Pathway: Global awareness and education is needed that new diseases usually come out of ruined ecosystems, and those environments are being ruined by our own dollar signals as consumers. Consumer economics thus drives the growing risk of pandemics—and equally offers a solution through informed consumers, ethical corporations and sustainable industries. Strengthen international efforts to restore soils, water, landscapes and oceans. Build price signals into food and other resource-based products that enable reinvestment of natural capital.

3. Establish worldwide early warning systems for new pandemics. Publicly fund a major global effort to develop new antibiotics and antivirals.

   Pathway: WHO and world medical authorities are already working on this. It needs to be coupled with predictive systems for ecosystems facing profound stress, whence new pathogens are likely to spread.

4. Destroy all stocks of extinct plagues. Outlaw the scientific development of novel pathogens with potential to harm humans.

   Pathway: like nuclear weapons, this pathway is blocked by the refusal of militarised nations to disarm. Only citizen and voter action can compel them.

5. Impose a code of ethics and public transparency on all scientific research—on pain of dismissal, refusal to publish and criminal penalties—with potential to create autonomous machine intelligence or robotic devices which take their own decisions to kill people.

   Pathway: it is time for all scientific disciplines to impose a code of ethics on their practitioners, to reduce the likelihood of science being used for evil or dangerous existentially risky purposes. Discussion at global scientific congresses should begin at once.

6. Establish a new human right to prohibit mass surveillance of entire populations and to restrict cradle-to-grave data collection on individuals not suspected of a crime.

   Pathway: Constitutional reform will be necessary in most cases to prevent governments, and stronger privacy laws to prevent corporations, from amassing data on all citizens and misusing it. Citizen and voter action will be essential to drive this. Transparency about, and public control over, data collection must become a fundamental pillar of democracy.

7. End poverty in all countries and redistribute human wealth more equitably as a primary requirement for the social cohesion necessary to preserve civilisation through its greatest challenges ever.

   Pathway: ending poverty is already cemented in global planning by the Sustainable Development Goals, however it is necessary to engage transnational corporations more fully in this task, since they now control most of the world's wealth. Dialogues around this have begun, but need to make swifter progress driven by awareness of the existential risk to all which disunity brings.

What You Can Do

- Live a more sustainable life. Select all your purchases wisely and thus share your wisdom through the potent influence of market economics.
- Practice the ancient human art of survival by anticipating risk: for every powerful new technology, ask yourself "What does this mean for my grandchildren?" and distinguish potential threats from opportunities.
- As a voter, demand laws which publicly disclose advances in artificial intelligence and nanoscience, so that there can be free and fair public

debate about which aspects of these powerful new technologies should be free and which should be restricted or banned.

- Take a moral stand against machines which can kill humans based on an autonomous decision.
- Take a moral stand against universal data collection and surveillance and their misuse. Demand constitutional reform to protect your freedom from spying.
- Understand that a fairer distribution of human wealth will lower the burden on the planet, increase the prospects of peace and plenty for all, and build the social cohesion necessary to counter major existential threats to civilisation and human existence. Support social justice as well as legal justice.
- Don't buy products or shares in companies that exploit and impoverish other people or damage the landscape, water or resources needed for human survival or who spy on their customers. Don't reward the wealthy for selfish behaviour.
- Require ethics, decency and fairness of all you whom deal with. Enforce them by your economic and democratic political choices.

9

The Self-Deceiver (*Homo delusus*)

The human brain is a complex organ with the wonderful power of enabling man to find reasons for continuing to believe whatever it is that he wants to believe

—*Voltaire.*

Over the years, a dozen aircraft have ploughed into the rugged flanks of Apex Mountain, British Columbia. While several causative factors were involved, a common thread is that pilots of low-flying aircraft found themselves trapped amid a landscape that is in fact higher and climbs more steeply than it appears to. "We use that particular terrain in one of our mountain courses to show our students the optical illusion. There is an appearance that the terrain climb is shallow but it's quite steep," local flying instructor Mark Holmes told the *Globe and Mail* (Theodore 2010; Youssef 2010). Canadian Transport Safety Board investigator Bill Yearwood added: "You can easily be lured into thinking that the terrain is not as high as it actually is," he said. "You reach a point where you can't turn around" (Theodore 2010).

The pilots probably died as a result of a false belief, engendered by an optical illusion, that their path ahead was safe. It's a not uncommon story through human history. Such a 'belief' is possibly what killed the young pre-human, who fell to the leopard 1.5 million years ago (Chap. 1) and many, many more since. On the grand scale, unsound beliefs could also prove fatal to civilisation.

If you ask them how they see, most people will say it's with their eyes. But of course, like many popular notions, this isn't true. Science has long known that our eyes consist of specialised nerve cells that collect and process

© Springer International Publishing Switzerland 2017
J. Cribb, *Surviving the 21st Century*, DOI 10.1007/978-3-319-41270-2_9

light, and that the actual image we see is in fact assembled in the brain from countless packets of information. Thus it is the brain which 'sees', not the eye. And, of course, the brain is fallible, subject to misinformation and heavily influenced by its past experiences when it constructs those images.

The mirage is a familiar optical illusion—an apparent sheet of water on an otherwise hot, dry road or desert landscape. The brain even 'knows' the water probably isn't there, and can verify this by approaching it—but it still insists on interpreting the light waves collected by the eyes as if it were real, probably because it is more accustomed to seeing water than shimmering air, or because the viewer is thirsty or else hallucinating. There are many charming tests that illustrate the ability of the brain to generate false images, or create illusory pictures—like the endless Escher staircase, the Kanisza triangle or the nineteenth century spinning disc depicting a bird in a cage.[1] These give rise to the familiar statement that "things aren't always as they appear". Our other senses are equally open to misinterpretation by the brain, though less spectacularly. Illusions are fun—but they can also be deadly. In the 1990s, for example, British researchers warned of an optical illusion that was probably killing around 50 children crossing the UK's roads every year: this was due to the way that drivers interpret the optical flow reaching their brain as their car moves rapidly along a road. The brain is accustomed to seeing adults crossing the road and this is what it expects to see, allowing the driver sufficient time to brake the car in normal circumstances. However, when the people seen are children, their smaller size means they are much closer than the driver's brain—accustomed to seeing adults—expects (Hamer 1994). The children thus die as a result of the driver's false belief.

These air and road accident cases are fair analogies for our current world and the way it is speeding towards major existential crises—even potential catastrophe—without humanity as a whole recognising the peril, probably because we are misled by past experience of a much safer, more stable, less overpopulated world and by our accumulated belief framework (which has achieved a hitherto highly-successful ascendancy of which we are inordinately proud) into underestimating or ignoring them.

The brain forms its constructs of the world around us not only from what our sensory organs tell it, but also from what it remembers from past experience or learns from received information. Where the sensory data is insufficient, rather than leave us floundering in the face of danger, the brain forms an unsubstantiated view, a best-guess or imaginary picture, of the situation

---

[1] A range of interesting optical illusions can be found at http://list25.com/25-incredible-optical-illusions/.

which allows us to fight, flee or take other action. That is a *belief*—a snap assessment founded on insufficient or no information, but which is often very strongly held on the basis of learned experience or training. Throughout our existence it has saved us many times: often, too, it errs and kills us.

Belief is a good thing, even an essential thing, because it enables us to navigate through life, exploit opportunities and avoid hazards quickly, without having perfect knowledge of them. Our brain simply fills in the blanks, erecting an image or construct that enables us to take rapid decisions and act with insufficient detail about what is actually going on around us. Belief can be compared with the CGI (computer generated imagery) used in epic fantasy movies like *Lord of the Rings*, where only a few elements or characters are 'real' and the surrounding scenery is wonderfully-contrived vision painted in with electrons and photons. Humans, having a more complex brain than other herd animals, may be able to erect more elaborate constructs—including the precious ability to envision future events and act in time to save ourselves. It is possibly this unique ability that differentiated us from other inhabitants of the African savannah a million or more years ago. We had a better picture in our minds of the world around us, how it worked, its dangers and how to deal with them that other social animals, even though its empirical basis was no more factual than that held by the wild dog, the baboon or the antelope.

In *Homo sapiens*, however, this simple form of belief has been elevated into something far more sophisticated. Something magnificent. Indeed, something which became central to our civilisation and its relationship with the world: organised belief, created out of our collective experience. This form of belief gave us the courage and the inspiration to cross new horizons, to open up new lands, test new ideas, experiment with novel social institutions and innovate technically, when our senses alone might have pinned us to the Palaeolithic. The Oxford English Dictionary defines belief as 'an acceptance that something exists or is true, especially … without proof' (Oxford Dictionaries 2016). Two examples: a king is no more powerful than any other individual, unless we choose to believe he is: it is our collective belief that confers his power, and the social order which derives from it. And a piece of gold has no more intrinsic value than a piece of copper, unless we choose to believe it does. There is no empirical proof for either case: it is the human mind than endows it with the 'reality' on which we act. From such simple consensual fictions is the fabric of our deeply complex modern society spun. Politics, religion, our monetary system, our social order, our loves, hates and prejudices, war and peace, our understanding of the ability of the Earth to support our needs, were all founded upon human belief and—far less often—on objectively testable, and re-testable, facts.

Aside from our fallible senses, external fact has only been imparted to belief by science in the past two or three centuries—and see how far and high the addition of such tested evidence to a worldview founded on belief has lofted humanity in so very short a time, a mere 0.03% of our existence. Folk in earlier times may have *believed* that people (like Icarus, witches or the Inca birdmen) could fly, but it took science, maths, metallurgy, engineering and a tested bed of evidence to get us airborne. The difference between modern civilisation and Roman civilization 2000 years earlier is that one of them was endowed with far more reliable facts about the nature of the world, leading to more reliable beliefs about it and thus, to better technologies for dealing with it, although both were equipped with elaborate political, religious and social structures. The main difference between the eighteenth and twenty-first century economies is that the latter is powered by science and based on tested facts.[2] And today, in the world scientific enterprise, we have built a mighty engine dedicated to the discovery of more evidence still—evidence which very often challenges or even overthrows our previous world-view. This state we commonly refer to as 'progress'.

In this book we have explored ten main categories of risk pivotal to the future of civilisation and our species. Each of these risks is founded on a growing mountain of objective evidence assembled over decades that can all be proved and independently re-checked. Unlike many aspects of human behaviour, none of these existential risks is founded upon belief alone: all are based upon things which can actually be measured and independently veri-fied. However so powerful is the habit of belief as an influence over human behaviour and politics that many people still trust it more than they trust the accumulating evidence before them (if, indeed, they are even aware of the detail of that evidence).

At this juncture in history, it is entirely possible that, by trusting too much to things which are unproven and unprovable, we fail to give due weight to more trustworthy proofs of the risks we face as a species. And thus we fail, as a species, to apprehend the clear and present dangers that surround us. In this way, belief, the mental process on which much of humanity's mar-vellous ascendancy is founded, becomes our Achilles' heel—the fatal flaw that brings us down when it should in fact uplift and preserve us. This is not because our beliefs are 'wrong': only that they provide us with imperfect

---

[2] Based on the 1987 Nobel Prize winning work of Robert Solow—http://www.nobelprize.org/nobel_prizes/economic-sciences/laureates/1987/press.html.

knowledge and understanding of our situation, and in an overcrowded, polluted and resource-depleted world in which major natural systems are starting to break down, they may no longer provide a sufficiently reliable framework to assess and ensure our future.

The other, essential, factor is that most human beliefs—especially those of older people—are founded upon an historical view of the world, our population, and the scale our physical demands based on the situation several decades ago, when they were young. Human numbers have recently become so huge that most people have simply not grasped how dramatically our world has changed in the space of less than a single lifetime—from one that was probably sustainable to one that evidently isn't.

In this chapter four main categories of belief are examined, for how they may help or hinder the cause of human survival. There is no intent to disparage or to criticise any belief or believer, but rather to illuminate how each may serve the greater good.

Money

The modern world is founded on a belief in money, a commodity that did not exist until about 5000 years ago and probably won't exist in the far future. Yet most people behave as if money were, in fact, real—rather than a consensual belief or a bond of trust between people. As Andrew Beattie puts it in Investopedia: "Money is valuable merely because everyone knows everyone else will accept it as a form of payment" (Beattie 2015). Yet many devote their lives to the 'making' of this insubstantial substance, and respect it almost as if it were some deity (which indeed it was, during Roman times).

Money is an urban idea created for convenience and efficiency—earlier hunter-gatherer and agrarian societies got along for tens of thousands of years without it, by gifting, exchange or barter of goods. Small silver bars were used in Mesopotamia's cities around 3000 BC, and a business code was drawn up by the Babylonian (Iraqi) king Hammurabi in 1760 BC to standardise rates of exchange between different goods, debt, contracts and fair business practice. The first metal coinage was produced by the Chinese Zhou dynasty in 1000 BC, followed by the kingdom of Lydia, in Turkey, in 600 BC (History of Money, Wikipedia, acc. 2015). Today, in an online electronic world, it is thought that as much as 90–95 % of the world's currency exists only in digital form (Grabianowski 2015; Swamy 2015), as mere electrons—not even as 'solid' atoms—in the computers of the banking and international exchange system. A large enough solar storm, akin to the events of 2012 and 1859

(Anthony 2014), could collapse the world's power grids and fry its computers with a vast magnetic shockwave, causing most of our money—or at least the records of it—to vanish in something even less substantial than a puff of smoke, scientists have theorised. Exchange would no doubt be re-established, but for a while there would be absolute chaos. However, while spectacular, even this is probably not a civilisation-threatening or species-threatening event—although it would be mightily inconvenient and would probably trigger second-round effects such as food panics, famines, disease outbreaks, energy crises and government failures.

The true risk to humanity from money consists of the fact that, being a creation of the human imagination it is, in theory, infinite. However, it is used to purchase, exhaust, pollute or destroy things which are finite—like soil, water, forests, fish, wildlife, certain minerals and energy sources, the Earth's climate.

A century ago, or even half a century, this simply didn't matter. A much smaller world population, living far more simply, could imagine as much money as they wanted to meet their needs. When their imaginations got a little overheated, the apparent risk ruptured the delicate public consensus about the value of things and a correction ensued—the Dutch tulip (Wood 2006) and South Sea stock (Encyclopaedia Brittanica) 'bubbles' burst, Wall Street crashed, and the Global Financial Crisis acted to bring money and other valuables back in line with the public understanding of their real value. As *The Economist* describes it:

> It is clear the [Global Financial] crisis had multiple causes. The most obvious is the financiers themselves—especially the irrationally exuberant Anglo-Saxon sort, who claimed to have found a way to banish risk when in fact they had simply lost track of it. Central bankers and other regulators also bear blame, for it was they who tolerated this folly. The macroeconomic backdrop was important, too. The "Great Moderation"—years of low inflation and stable growth—fostered complacency and risk-taking. A "savings glut" in Asia pushed down global interest rates. Some research also implicates European banks, which borrowed greedily in American money markets before the crisis and used the funds to buy dodgy securities. All these factors came together to foster a surge of debt in what seemed to have become a less risky world (The Economist 2013).

The GFC began with "a flood of irresponsible mortgage lending in America", states *The Economist*. Banks and lenders simply plucked funds out of thin air to lend to homebuyers who couldn't then repay their 'debt', especially once house prices began to slide. These dud loans (made from imaginary money) were then bundled into derivatives and on-sold to people who didn't know what they were buying, and when confidence in their value deflated, it unleashed a chain reaction of failures. Ironically, the crisis was gradually solved by the US Federal Reserve doing pretty much the same thing, creating floods of new money out of thin air to pump up punctured confidence and bail out the banks, and dressing up this pecuniary prestidigitation in the fancy term 'quantitative easing'. An additional US$3.5 trillion was 'imagined' into existence by the American monetary authorities in this way: "So the $US3.5 trillion the government pumped into the economy was 'created' from nowhere. But that money didn't just go into the pockets of everyday consumers. It was used to buy the banking industry's debts. The money went into the banks' reserves which strengthened banks and gave them the confidence to loan money, which allows the economy to chug on. The alternative was to let the banks crash which could set off deflation, which is essentially a contraction of the economy," News Ltd reported (News Ltd. 2014). Thus, money that did not previously exist was used to cancel the imaginary debt created by imaginary money; no wonder the public was confused about how it all happened.

In terms of the issues which challenge the survival of civilisation, and possibly the species, one of the chief ones is the fact that money—despite the best efforts of the central banks to control its supply—is in theory infinite, whereas it is often used to extract, exploit or damage things which are finite, like soil, water, air, energy, metals, fertiliser, wildlife and timber. It is this mismatch between a theoretically infinite purchasing power and an increasingly limited supply of material goods, brought on by overpopulation and unrestrained demand, that is responsible for the damage which human economic activities are now wreaking on the Earth's natural systems—climate, the biosphere, our health. Money is the medium by which the demand for improved lifestyles of a superabundant humanity is now treading ever more heavily upon a finite planet. Money sends the signals which drive us to overuse, overexploit and pollute.

Classical economists argue that scarcity tends to push up the price of goods that are scarce, thus rationing them and creating alternatives—but one of the blind-spots of economics is its tendency to 'externalise' (ie not account for, or ignore) the true costs of things like soil degradation, pollution of air and water, loss of species, landscape degradation, extinction of the Holocene climate, warfare, refugee tsunamis, pandemics and the poisoning of future generations of young humans. Economics simply puts these things aside, as if they did not exist. Yet all of these risks are driven, at core, by monetary signals. The classic example of an 'externality' is the 'Tragedy of the Commons', where unrestricted human demands place too great a burden on a limited, shared amount of land (or water) and end up ruining it—an event which many commentators consider is now taking place at planetary level. These events are extraordinarily hard to rectify—because the market fails to self-correct them, warn Dasgupta and Ehrlich (Dasgupta and Ehrlich 2013). We spend, but we do not count the cost.

A paradox is that we humans are intensely competitive beings—but at the same time intensely collaborative. Both are essential aspects to our being. Indeed, almost nothing (including competition) is achieved without collaboration. Our early survival and ascendancy was probably due more to our collaborative nature than to our competitive character. However, in the past 200 years, with the rise of money as a ruling factor in societies (that were previously largely moneyless) the competitive element has become predominant and is indeed now the chief driving factor in the global economy, including its largest players, the US and China. As we have seen, this is leading to a widening gap between rich and poor in all societies, which could sap the cooperation essential to our collective survival. It is therefore of overwhelming importance that we humans re-balance our behaviours and beliefs, between competing and collaborating. How this might occur is addressed in the next chapter.

Thus the popular belief that 'money is the main aim in life' stands between humanity and our ability to deal with and overcome actual threats to our future existence. As long as monetary signals instruct us to overexploit, destroy, degrade, compete and damage the place we live, it will be extremely hard to put in place other systems which preserve, protect, restore, recycle and reuse. Common sense and wisdom tell us to protect and preserve the Earth on which we live, and our own future as a species: money, looking short-term, commands the opposite. In short, modern humanity's faith in money is presently more powerful than our instinct for self-preservation. That is a worry.

It is all too easy to dismiss this as 'greed' or to blame 'capitalism', as some do: laying blame and wrangling over politico-economic philosophies hasn't provided lasting solutions to most human problems in the past, and abolishing money or our monetary system won't solve anything, even if society were inclined to do so—which, so far, it isn't. In her book *This Changes Everything: Capitalism vs. the Climate* (Klein 2014). Naomi Klein argues for an alternative approach of consciously putting climate ahead of money. She says that climate change "if treated as a true planetary emergency," could "become a galvanizing force for humanity, leaving us all not just safer from extreme weather, but with societies that are safer and fairer in all kinds of other ways as well." Well, maybe, but what are the chances if big global monetary signals, and society's love of money, continue to point the opposite way? And what of the nine other existential threats?

Coming from a slightly different perspective US academic Jeffery Sachs argues for a reconnection between our economic system and our fundamental need to survive: "By separating nature from economics, we have walked blindly into tragedy," he says (Sachs 2015b).

> We have entered a new age of sustainable development whether we like it or not, even whether we recognise it widely or not. As the great biologist E O Wilson has put it, we have stumbled into the 21st century with stone-age emotions, medieval institutions, and near godlike technologies. In short, we are not yet ready for the world we have made. The sustainable development goals will be a vital opportunity to give ourselves new guideposts and measuring posts for prosperity, justice, and environmental safety in our fast-moving, rapidly changing, and dangerously unstable world.

In a third perspective, the International Resource Panel (cited in Chap. 3) argues for 'dematerialisation' of the world economy—that is, we need to build an economy that is much less dependent on material things for its growth. An economy that uses less energy, less metals, timber, wild animals, less soil, less concrete etc to generate growth, and which creates more of its growth and employment from products of the human mind, imagination and creative impulse (International Resources Panel 2011). To begin with, this can be achieved by the simple and economically-attractive course of improving production efficiency—in agriculture, mineral production, energy use, manufacturing, transport, construction. When that has been carried as far as it can be, the goal then becomes the transformation of the economy into a system in which 'wealth' is predicated mainly on knowledge and creativity, rather than on material assets and products.

Surely it makes more sense to invest a theoretically infinite supply of 'wealth' in an equally infinite commodity—human knowledge and imagination—than to use it to degrade and destroy the finite things we need for our survival? To de-couple 'money' from material goods and put it to work on the 'economy of the mind'—science, philosophy, art, literature, entertainment, information, software, design, technology, sport, fashion, cuisine, health and caring for fellow humans. These can all be performed with modest material input, and most of that, recycled or renewable via the 'circular economy'.

With robots performing most of the dirty, boring and arduous tasks in the material economy by mid-century the opportunity to de-couple the material from the immaterial economies will be even greater, to concentrate most of our growth, prosperity and employment in the latter, rather than the former. The old argument that "we gotta cut down the forest to protect forestry jobs" will no longer hold sway, because most of the jobs will be in the immaterial economy. Since we will still need to exchange material goods and stimulate their reprocessing and re-use, it may be expedient to develop a second "currency", but one with a fixed supply and value, which is immune from the distortions of the feckless speculators and gamblers in the banks, the trading rooms and the dodgy loans departments of the world. A finite monetary system for a finite world. An 'Earth Standard' currency (not unlike the old gold standard) that reflects the true scarcity of material resources and signals as they become dangerously scant and which helps allocate them better. Wiser minds will suggest other solutions to the financial challenge that confronts us. This is purely a contribution to the debate we have to have.

However, the one absolutely essential understanding is that we must not permit our wonderful, illusory commodity, money, to misguide us into turning the real Earth into a hot, toxic, uninhabitable slag-heap.

Politics

For more than 3000 years, humans have argued over politics (the "affairs of the city" in ancient Greek) and governance structures. Monarchy, theocracy, oligarchy, timocracy, tyranny, feudalism, democracy, republicanism, parliamentary democracy, socialism, capitalism, Marxism, militarism, fascism, central planning, libertarianism and their many variants have each been espoused as the ideal way to run a society—and tens of millions have perished, pointlessly, to try to prove the superiority of one theory over the others.

2400 years ago, Plato inclined to the notion of a wisdom-loving autocrat, could we but find one:

> There will be no end to the troubles of states or indeed, my dear Glaucon, of humanity itself, till philosophers become kings in this world, or till those we now call kings and rulers really and truly become philosophers, and political power and philosophy thus come into the same hands (Plato. C 360BCE).

Of the principal political alternative, a couple of millennia later, an electorally-defeated Winston Churchill lamented "Democracy is the worst form of government, except for all those other forms that have been tried from time to time" (Churchill 1947). If philosopher kings are in short supply, today's autocracies more obsessed with building their wealth, power and prestige than with saving humanity and today's democracies can no longer make their minds up about anything truly important to the long-term future of themselves or the species, whence is the wisdom for human survival to come?

Probably not from politics, even democracy, in its current guise. The disenchantment seeping through western democracy was highlighted in the late Peter Mair's chilling analysis *Ruling the Void: The Hollowing of Western Democracy*, in which he observed: "The age of party democracy has passed. Although the parties themselves remain, they have become so disconnected from the wider society, and pursue a form of competition that is so lacking in meaning, that they no longer seem capable of sustaining democracy in its present form" (Mair 2014). As a result, electoral turnouts are in decline, membership is shrinking in the major parties, and those who remain loyal partisans are sapped of enthusiasm and retreat into dogma. Extremism is once more on the rise.

A reason for this is that, slowly but surely, all national governments—democratic or autocratic—are losing their authority over their own economies and political affairs, as the demands of a globalised world grow ever more imperious. International agreements and treaties, supra-national bodies like the G8 and G20, the EU, the OECD, ASEAN, the World Trade Organization, the International Monetary Fund, the United Nations and WHO are gathering influence, while national governments appear manifestly inadequate, or else disinterested, in dealing with planet-scale challenges such as climate change, resource scarcity, species extinction, population and global poisoning. Suffering this power haemorrhage, national politicians in quest of electoral success are increasingly led to lie or exaggerate about what they can really achieve for their constituents. And voters become more sceptical and disengaged as a consequence.

At the same time world economic power is concentrating in the hands of global corporations who are adept at avoiding taxes—so draining governments of their lifeblood—or regulations they don't like and who, as a general rule, display limited interest in matters pertaining to human survival. The Global Policy Forum has noted "Of the 100 largest economies in the world, 51 are now global corporations; only 49 are countries. Wal-Mart—the number 12 corporation—is bigger than 161 countries, including Israel, Poland, and Greece. Mitsubishi is larger than the fourth most populous nation on earth: Indonesia. General Motors is bigger than Denmark. Ford is bigger than South Africa. Toyota is bigger than Norway" (Anderson and Cavanagh 2000). The top 200 corporates had more economic might than the bottom 80% of humanity, it added.[3]

None of these giant corporations is a democracy. Each is a functioning autocracy, operating a top-down, command-and-control model. Some corporates may acknowledge it—the parlance is corporate social responsibil-

---

[3] Fort a recent list of the 'top 200' see: Forbes 2014.

ity (CSR)—but none are explicitly there for 'the public good'. All are more devoted to the pursuit of the immaterial dollar than they are to preserving civilisation. Few are managed by 'philosopher kings' who value wisdom, but mainly by individuals motivated by vast salaries, shares and bonuses into often unwise, short-term, rent-seeking behaviour (see Chap. 5). The combined power of these 40,000 transnational companies is emasculating national governments, not only sapping their revenues, but also their legal authority, public confidence, freedom of action and ability to decide what happens within their own borders and make it stick. The logical outcome of this process is that by mid-century most governments—democracies especially—will find themselves increasingly paralysed, discredited, broke, unable to govern, powerless to deliver to their citizens the benefits they promised when elected. Globally-connected megacities will arise as more effectual governmental entities, most of them far more capable of delivering services like schools, hospitals, roads, energy, clean water and amenities that enhance the lives of their citizens. Thus, with power concentrating at both global and local levels, the twenty-first century may well witness the hollowing out, if not the actual demise, of the nation state as a political construct (see for example, Khanna 2013). And good riddance, since national governments and nationalism are manifestly a cause of most wars—and are the main proponents of weapons of mass destruction. If humanity is to be destroyed, it will be done by nations.

The peril which this evolving situation imposes on the human future is that many people will remain distracted by the delusion that some form of state political structure or ideology offers the magical solution to the existential threats that confront us. They don't. Not one of them. Humanity is not going to be rescued by Marxism or by capitalism, by Westminster democracy or autocracy, by free trade or protectionism, liberalism or conservatism: these are eighteenth, nineteenth and twentieth century ideas whose relevance has been subsumed by the massive burgeoning in human numbers and demands. In future, survival and prosperity will depend on humans and their cities co-operating across the world for our collective good. If we allow ourselves to be distracted by time-worn debates over political 'isms', and divided by archaic ideologies, the risk is that, like those Canadian pilots, we fly further and further up a steepening valley, trusting to the enduring illusion that politics and its rituals will deliver us, blind to the adamantine walls of reality that are closing in around us.

The survival of civilization and the human species must become the pivotal ambition of *every* political belief, system, party and representative, their first policy goal. Without such consensus, politics will serve only as a dangerous diversion. But politics will only change if we, the people, compel it to refocus on what is most important to our future. In the twenty-first century, the political party that does not place human survival as the central plank of its policy platform is not worth even a single vote. That is our loyal duty as electors, or as party adherents.

At the same time transnational corporates, the new economic superpowers, will need to stop rewarding their CEOs for wrecking the planet on the basis that, if nothing else, it's bad for business. How they might do that will emerge in the next chapter.

Religion

Since the shamanism of the Stone Age, religious belief has been the primary construct on which humanity has founded its vision of the world, its moral laws and social order. Like all forms of belief on which we rely, faith provides a way of interpreting and dealing with a world we do not fully know or comprehend but nevertheless must navigate. It is likely to be as significant a power and influence over human affairs in the twenty-first century as in the past.

A worldwide opinion poll of 50,000 people found in 2012 that 59 % of respondents had a religious faith and 36 % did not (WIN-Gallup International 2012). It noted a 9 % drop in religiosity over the previous 7 years. By 2050, the Pew Research Center forecasts, the two largest religions—Christianity and Islam—will have 2.9 billion and 2.8 billion adherents respectively, and Hinduism 1.4 billion (Pew Research Centre 2015). Religious growth, it noted, will depend chiefly on where the highest birth rates occur.

Religious faith has proven both a great strength and sometimes a fatal weakness for humans. Many faiths, while asserting their own truth, have a habit of denying the truths of others, and this often ends in tears. Between 1618 and 1648, for example, Europe was plunged into one of the bloodiest and most brutal sectarian conflicts in its history, between Catholic and Protestant states of the fragmenting Holy Roman Empire. It caused famines and epidemics, killed 7.5 million people, bankrupted many countries and spawned obscenities like the sack of Magdeburg (Necrometrics 2012). The Taiping Rebellion in China from 1850 to 1864, even more bloody, was propelled by religious (Christian) visionaries and is thought to have cost 20 million lives. The guesstimate for the loss of native Americans who died as a result of a religion-and-greed-inspired European invasion is 17–30 million (Encyclopaedia Britannica, acc 2015). The Sunni-Shia sectarian schism in Islam has claimed uncounted lives in a conflict that has rumbled on for nearly 14 centuries (Council on Foreign Relations. Acc 2015). Overall, religious conflicts have probably destroyed somewhat fewer lives than politically-inspired wars through history; both nevertheless represent powerful proof of what happens when human belief systems collide—even those which appear to the outsider to exhibit only trivial distinctions of faith, doctrine or worldview.

The danger this poses to a humanity facing great existential challenges is plain: we risk spending more time and energy disputing the details of what we believe than we do in ensuring there is anyone left to believe in anything.

The history of religion teaches us that no faith has a monopoly on truth, no matter how certain—or fanatical—its followers. Almost every faith contradicts the beliefs of every other faith. Indeed, if every individual were to be closely examined on what they believe it is doubtful if any two, even of identical creed, would subscribe to *exactly* the same beliefs: having different brains and different life experiences, each person holds a slightly varying personal concept of deity, spirit, afterlife or morality, for example. Yet this disparity does not weaken nor invalidate religion: rather, though diversity, it strengthens it. As with politics, the essential task is to focus the world's religions, greater and lesser, on the common cause of human survival and on the shared convictions which serve it—rather than on doctrinal distinctions and disputes. "There will be no peace among the nations without peace among the religions. There will be no peace among the religions without dialogue among the religions" is the proposition put by Dr Hans Küng, a German Professor of Ecumenical Theology, who supports a 'Parliament of Religions' and has campaigned for a 'global economic ethic' (Musser and Sunderland 2005). However, unity among the faiths must now embrace issues far wider than the traditional goals of world peace, political and economic equity. It must also embrace the realities of the end of the climate in which humans arose, the extermination of half the world's living creatures, the draining of its resources and the loss of its forests, grasslands and soils, the poisoning of its children, air, water and food, and the relentless march of technologies and businesses ungoverned by the very morality and ethics that, often through religion, enabled human society to succeed in the past.

This in turn may call for revision of certain ancient articles of faith: the belief that divinity will save us no matter what we do, if we just ask penitently and sincerely enough. The belief that humans are 'lords of creation' set over the Earth to do as we like with it. The belief that we should have as many children as possible, regardless of the hazards and misery caused by overpopulation. The belief that the deity rewards piety with material riches. These are examples of past beliefs, perhaps once useful, which in the contemporary world and in the light of scientific evidence about our jeopardy, are now clearly detrimental to our chances of survival.

The worlds of science and religion are often depicted as opposed to one another—yet in reality both are sturdy components of the same broad belief system which humans have used for millennia to navigate, survive and prosper in a challenging universe. Albert Einstein, who probably knew more about the cosmos than any other person of his day, found little difficulty in reconciling his views as a physicist with his beliefs as a person of faith, famously stating "Science without religion is lame, religion without science is blind" (Einstein 1939). The partnering of faith and science has yielded the enormous attainments of our present Age—as well as giving rise to the enormous perils which now encompass us. One operates on the accumulated experience of humanity and its evolutionary predecessors over tens of millions of years about what is good and bad for us, which we know as morality (Harris 2013), the other operates on impartial facts it can test and theories it can validate or invalidate about how the world works. Neither provides 20:20 foresight: religious people seek divine guidance, scientists use computer models. However, like our legs, these two integral components of our belief system about our world now support humanity and carry us forward. Amputating one of them would not be wise.

A significant development in this field in recent times was the encyclical *Laudato Si*, issued by the head of the Roman Catholic Church, Pope Francis in 2015, in which he said that the Earth

cries out to us because of the harm we have inflicted on her by our irresponsible use and abuse of the goods with which God has endowed her. We have come to see ourselves as her lords and masters, entitled to plunder her at will. The violence present in our hearts, wounded by sin, is also reflected in the symptoms of sickness evident in the soil, in the water, in the air and in all forms of life. This is why the earth herself, burdened and laid waste, is among the most abandoned and maltreated of our poor; she "groans in travail" (Rom 8:22). We have forgotten that we ourselves are dust of the earth (cf. Gen 2:7); our very bodies are made up of her elements, we breathe her air and we receive life and refreshment from her waters.

In a powerful linking of science with theology, Pope Francis inveighed against global pollution and our 'throwaway culture', explaining that "The climate is a common good, belonging to all and meant for all. At the global level, it is a complex system linked to many of the essential conditions for human life. A very solid scientific consensus indicates that we are presently witnessing a disturbing warming of the climatic system." He warned also about the dangers of water scarcity, the extinction of species, the damage to the oceans, declines in the quality of human life, inequality and the 'breakdown of society'. He also called the causes, stating "It is remarkable how weak international political responses have been. The failure of global summits on the environment make it plain that our politics are subject to technology and finance." In conclusion, he urged Catholics and others:

> We must regain the conviction that we need one another, that we have a shared responsibility for others and the world, and that being good and decent are worth it. …. When the foundations of social life are corroded, what ensues are battles over conflicting interests, new forms of violence and brutality, and obstacles to the growth of a genuine culture of care for the environment (Pope Francis 2015).

Although Catholics make up barely a sixth of the world population there is no doubt that the Pope's example, knitting science with belief, has given other major religions fresh inspiration. Shared views among the world's great religions on climate in particular had been building for some time. In 2009 Hindu leaders, representing another sixth of the Earth's population, stated "Hindus recognize that it may be too late to avert drastic climate change. Thus, in the spirit of *vasudhaiva kutumbakam*, "the whole world is one family," Hindus encourage the world to be prepared to respond with compassion to such calamitous challenges as population displacement, food and water shortage, catastrophic weather and rampant disease." Islamic scholar Hyder Ihsan Mahasneh, asked by the Moslem World League to compile a faith statement on the environment, found "several Qur'anic principles… taken in their totality… state in clear terms that Allah, the One True God is the Universal God and the Creator of the Universe and indeed, the Owner of the Universe. To Him belong all the animate and inanimate objects, all of whom should or do submit themselves to Him."

> If biologists believe that humans are the greatest agents of ecological change on the surface of the earth, is it not humans who, drawn from the brink, will—for their own good—abandon Mammon and listen to the prescriptions of God on the conservation of their environment and the environment of all the creatures on earth? The Islamic answer to this question is decisively in the affirmative," he concluded (Mahasneh 2003).

In an *Islamic Declaration on Global Climate Change* a group of academics, Muslim scholars and international environment policy experts called on the world's 1.6 billion Muslims to phase out greenhouse-gas emissions from fossil fuels and switch to energy from renewable sources. Unlike Roman Catholicism, Islam has no central religious authority but, according to the journal *Nature,* "the declaration suggests Muslims have a religious duty to tackle climate change" (Castelvecchi et al. 2015).

Church of England Bishop of Sheffield Steven Croft articulated the peril of climate change as "a giant evil; a great demon of our day", adding "Its power is fed by greed, blindness and complacency in the present generation, and we know that this giant wreaks havoc though the immense power of the weather systems, which are themselves unpredictable," (Jones 2014). The Church called for an 80 % cut in Britain's emissions by 2050 and began to sell off its own investments in fossil fuels. In Australia, Anglicans joined with Hindu, Bhuddist, Uniting Church, Jewish and Catholic faiths in a letter urging stronger action by their Government to cut carbon emissions. In the USA, 380 American Rabbis signed a similar public appeal:

> The Torah warns us that if we refuse to let the Earth rest, it will "rest" anyway, despite us and upon us – through drought and famine and exile that turn an entire people into refugees. This ancient warning heard by one indigenous people in one slender land has now become a crisis of our planet as a whole and of the entire human species. Human behavior that overworks the Earth – especially the overburning of fossil fuels – crests in a systemic planetary response that endangers human communities and many other life-forms as well (Waskow 2015).

Thus, all of the world's great faiths are approaching a consensus, founded on science, on what is necessary for human survival in the context of climate change and broader environmental damage. If civilisation is to be preserved and the manifest threats to our species abated, humanity needs a stronger, sounder belief architecture on which to base its actions—not a weaker. As Einstein and Francis both suggest, religion + science is the way forward: it isn't a case of either/or. Science can provide religion with trustworthy facts about the world we inhabit to validate and strengthen our beliefs about it; religion can provide science with the moral compass that points to what is good or bad for humanity and our world.

Because science is morally neutral, without guidance from religious or other social values, species-threatening technologies such as weapons of mass destruction, the poisoning or degradation of the planet, uncontrollable artificial intelligence or other technologies (Chap. 8) are liable to proliferate before we awaken to their consequences. Just as we need to set physical boundaries to the damage we inflict on the Earth—as Rockstrom and Steffen have wisely proposed in their two ground-breaking scientific papers (Steffen et al. 2009; Rockström et al. 2009)—we also need to define moral boundaries to the way we use, overuse and misuse technology and other life. For all the university ethics committees, peer review and personal standards of individual researchers there is little moral governance of science, which is chiefly driven by commercial, defence or political funding imperatives or else by an unguarded fascination with opening Pandora's Box—even if ills then escape. This has led to the release, mass production, overuse and misuse of technologies which together now manifestly imperil the human future—besides offering a multitude of short-term benefits. Science is the primary driver of our present plight and cannot escape the moral responsibility for it. Therefore, it is in striking a balance between the benefits of science and technology and their unwise or over-use that a religion/science partnership has most to contribute.

The Human Narrative

The fourth field of potentially lethal human self-delusion may seem trivial compared with money, politics and religion, but is not. It is about the narratives we tell about ourselves—and about how they can lead us to misread true risks.

Hollywood over many years has terrified and delighted its audiences with a diet of epic disaster movies: *The Day After Tomorrow, Deep Impact, The Road, Independence Day, The Day After, On the Beach, 2012, The Stand* are all classic examples of the spinechilling cinema genre which envisages various gruesome end-of-the-world scenarios (IMDB 2012). The storylines mostly have one thing in common: civilisation is brought low but, against all the odds, the heroes survive. They, if not most of the world, are saved.

It's pure fantasy. We know from the history of cataclysmic wars and natural disasters of recent centuries that most heroes don't survive, that countless good and innocent people perish, that good does not always triumph over evil: or, as the songwriter Leonard Cohen succinctly put it, 'everybody knows the good guys lost' (Cohen 1988). Yet still we solace ourselves with the myth. The storylines of innumerable movies, novels and computer games provide the happy endings and diet of winners which their producers know will tap a stream of that invaluable faith-based commodity, money.

Computer games are a particular concern. Every night, around the world, a billion children go online to massacre sundry foes, villains and monsters in cyberspace in a plethora of artistically violent games. Children have always played at war—but in the cyber world, their victims are electrons and have no meaning, there is no penalty or personal risk from injuring or killing someone else; indeed, most games teach the player that the more they kill, the greater the reward—whereas a child who injures a real child is usually punished. An emerging danger is that young humans become mentally hardened to meaningless slaughter, unable to empathise; arguably, a generation is now in training to become psychopaths (Grodzinski 2011). Anders Breivik, the Norwegian gunman who slew 77 teenagers and passers-by in a bloody massacre in 2011 gave testimony that he trained for his assault by playing the first-person shooter video game *Call of Duty: Modern Warfare 2* (Harvard Kennedy School 2015). Breivik also sought to justify his rampage in a violent (and confused) manifesto against Islam, Marxism, multiculturalism and feminism, and pro-Israel, illustrating a weird interplay of political, religious, nationalistic and cultural beliefs—and their potential for tragedy when woven together in a fantasist mind (Hartman 2011).

Whether video games engender increased mass violence or terrorism is debated by sociologists (see for example, Markey et al. 2014), but the issue here is about belief: if billions of young humans are trained to believe they can kill, repeatedly, and get away with it, if they train against foes who are de-humanised by the fantasy cyber-medium they inhabit, then such beliefs are liable from time to time to blur or over-ride the morality and human relationships which ordinarily keep society in balance. Unlimited mental violence becomes a part of human social conditioning, part of our common narrative that then traps us within an endless cycle of actual violence. The deliberate recruitment by various militaries of these young 'Nintendo warriors' to pilot the actual robotic killing machines now being deployed on battlefields around the world is a case of how fantasy entertainment has been cynically turned into reality. Ironically, an increasing incidence of stress disorders and resignations among these drone pilots has highlighted the guilt, angst and self-disgust which some humans feel when compelled to perform constant, merciless killing (Chatterjee 2015). "How many women and children have you seen incinerated by a Hellfire missile? How many men have you seen crawl across a field, trying to make it to the nearest compound for help while bleeding out from severed legs?" Heather Linebaugh, a former drone imagery analyst, wrote in *The Guardian*. "When you are exposed to it over and over again it becomes like a small video, embedded in your head, forever on repeat, causing psychological pain and suffering" (Linebaugh 2013).

For humanity to come to a wise appreciation of the existential risks we run, it is essential we learn to distinguish better between reality and delusion at all levels. For thousands of years we have told each other stories about our heroes, their cunning, skill, bravery and their fate, to reinforce the beliefs we hold about ourselves. In the original age of myths and legends, those stories often echoed real life: Greek heroes frequently met with tragedy and divine punishment, the Norse Gods ended in Ragnarok, the Egyptian gods weighed your heart against the feather of truth. These stories held both moral force and important life lessons. Today's most popular legends however involve the hero—with whom we are invited to identify—always winning, never dying and often using mass violence to gain their reward.

These are the attributes of a semi-delusional humanity, one that cannot read the world or others with verity, one with diminished prospects for survival. Just as politics, religion and economics must recalibrate their philosophies and teachings to acknowledge the primacy of the need to work together for survival, the industry that spins our popular narratives—and its creators worldwide—must recognise the need for a new kind of hero for our uniquely perilous age, a hero who builds, co-operates, tolerates, embraces, cleanses, heals, restores, nurtures and sustains to gain a safer world. A hero more like a female, than the violent masculine caricature that darkens the present human storyline (see Chap. 10).

Decline or Crash?

Can human numbers continue to expand unchecked, until there is one person standing on every square metre of dry land on the planet? This chilling scenario was once depicted by the agricultural scientist Derek Tribe in *Feeding and Greening the World* (Tribe 1994) to illustrate the sheer absurdity of the argument for untrammelled population growth. Anyone who has devoted even a few seconds thought to the issue and read the UN population forecasts, will realise its utter impossibility. At some point this century the world is bound to experience 'peak people', and the only question is whether the subsequent decline in our numbers will be gradual, managed and consensual— or catastrophic, like most other biological collapses.

Catastrophic population crashes happen all the time in nature. Locust plagues boom—and then bust when their plant food runs out. Mouse and lemming plagues burgeon in summer—then starve in winter. Recent research indicates Europeans suffered a 60 % crash in numbers due to the Black Death in the C14th (Benedictow 2005). Between 1618 and 1680 around half the world population died prematurely in an interconnected series of famines, wars and disasters (Parker 2013). In the past 70 years, scientists have recorded no fewer than 727 mass mortality events (ie events where over 90 % of a given population perished) among fish, birds and mammals (Fey et al. 2015). In a chillingly apposite case, 29 reindeer introduced to St Matthew Island off Alaska in 1944 quickly expanded to more than 6000, ran out of food, then collapsed back to 42 animals (Klein 1966)—which is a fair analogy for what could happen to us on our island planet, if we fail to heed the alarm bells.

There is nothing in the whole scientific literature of biology to suggest humans are immune from such a crash in our numbers, whatever individuals may choose to hope or believe. If we break nature's basic rules for survival sooner or later, like all other species, we pay the price. This is the issue which Paul and Anne Ehrlich first sought to bring to our attention in 1968 with *The Population Bomb* (Ehrlich 1968). It is the fundamental admonition which the Club of Rome placed before an incredulous humanity in *Limits to Growth*, its ground-breaking 1972 computer prediction of the combined effects of exponential economic and population growth in a world of finite resources (Club of Rome 1972). It is the warning uttered in various guises by William Catton in *Overshoot* (Catton 1982), 'Sonny' Ramphal in *Our Country, The Planet* (Ramphal 1992), Tim Flannery in *The Future Eaters* (Flannery 2002), Lester Brown in *Outgrowing the Earth* (Brown 2004) and the lesson Jared Diamond extracted from the disasters of past civilisations in his masterwork *Collapse* (Diamond 2006), along with innumerable other respected authors and scientists. It is the reality which the Global Footprint Network has striven to illuminate since 2003, when it formed a partnership dedicated to "a sustainable future where all people have the opportunity to live satisfying lives within the means of one planet" (Global Footprint Network 2016). It formed a significant theme in Pope Francis's 2015 encyclical letter *Laudato Si* (Pope Francis 2015). It has been repeatedly dismissed as 'Malthusian nonsense' by boosters,

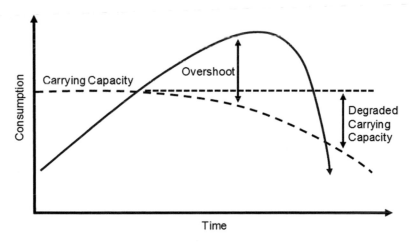

Fig. 9.1   How the Earth's ability to support us declines more rapidly as human numbers increase

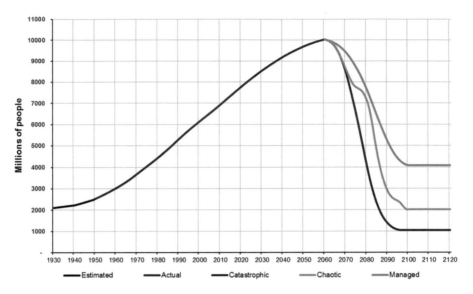

Fig. 9.2 Comparison of three scenarios for human population decline—catastrophic, chaotic and managed. The choice is ours to make

religious minorities, some sociologists and technophiles (who believe technology or blind faith will somehow solve all our problems, apparently without understanding them, or much collective effort) (Lomborg 2001).

The warnings of all these wise humans—and many more like them—are captured in Figs. 9.1 and 9.2.

Figure 9.1 illustrates the bind that we humans have got ourselves into and how it is likely to play out through the twenty-first century. Essentially, as our numbers and resource demands exceed the 'carrying capacity' (or renewal ability) of the Earth system, its ability to support us falls away, which in turn precipitates a steepening collapse in human numbers. One that will be both unimaginably horrible and extremely painful for those who have to undergo it.

Figure 9.2 shows the distinction between (1) a catastrophic 'natural' reduction in numbers (a human population spike, in which our numbers soar to 10–11 billion before crashing again to a billion or fewer) due to a combination of unsolved existential crises, (2) a 'chaotic decline' in which civilisation more or less disintegrates in the face of a crescendo of lesser crises involving climate, wars, pandemics, local eco-collapses and famines but manages to stabilise our numbers at a much lower level, perhaps two or three billion, and (3) a 'managed reduction' in which we go into overdraft on the Earth's resources but through wise choices succeed in managing our population back down to a sustainable 2–4 billion, enjoying a uniformly good quality of life, by the early twenty-second century—without destroying the biosphere or precipitating nuclear war and by recycling of resources, repair of ecosystems, arresting global warming, dematerialising the economy and reducing our fertility.

It may seem to the individual that she or he has little influence over which of these three scenarios prevails. This is not true. We can all affect the outcome—and the fate of humankind. How we do this depends to a very great extent on what we believe—since belief is such a powerful motivator of our actions—and on whether those beliefs are founded on a bedrock of scientifically testable fact and sound morality, or upon an treacherous mire of conflicting human imaginings and lusts. For survival, a wise humanity will be guided by beliefs that are both well-evidenced and moral.

What We Must Do

1. Reframe our economic, political, religious and narrative discourses to place the survival of civilisation and the human species as unquestionably the primary goal.

   Pathway: this requires worldwide action by leaders, the media, teachers, priests and concerned citizens. The evidence in this book underpins the arguments they can use.

2. Dematerialise the world economy to prevent the overuse and destruction of limited resources such as air, water, soil, forests and wildlife.

   Pathway: this is outlined by the UN Environment Program's International Resources Panel. It should be coupled with new thinking about industrial ecology, green manufacturing, recycling and growth of the creative economy—subsidised, if need be, by governments to transfer jobs from the 'old' to the 'new' economies as smoothly as possible.

3. Encourage the world's religions to take more courageous moral leadership over issues that threaten humanity as a whole, and to set aside their differences for the common weal.

   Pathway: religious leaders of all denominations must embrace the future, and the challenges it presents, not merely cling to the past—or risk being seen by society as redundant. They must focus on the evidence and what it says about our situation, as well as the moral good.

4. Refocus world, national *and* local politics on human survival through citizen pressure.

   Pathway: politicians are gradually learning that the path to political survival lies in attending to their concerns of their voters about human survival. However, all politicians need to undergo a short education in existential risk, as many still discount it or lack good information about it.

5. Inspire a new narrative about humans which prizes co-operation, tolerance, restoration, cleansing and sustaining over older, more divisive, selfish and destructive values.

   Pathway: in a way this revisits the more co-operative, altruistic movements and belief systems of the past, rather than the savage legacy of the twentieth century. People are ready for a happier, less murderous and exploitative human story—and it is the great opportunity of storytellers to tell it.

What You Can Do

- Appreciate that money, politics, religion and entertainment are beliefs and need to be reinforced with reliable facts to be of greatest service to us.
- Be extremely careful how you spend your imaginary dollar. It has very real consequences.
- Buy from, invest in and work for only companies that value the long-term survival of you and your grandkids.
- Vote only for only political parties and leaders who value the survival of you and your grandkids.
- If you are a person of faith, do whatever is possible to lead your co-believers to a position where the survival of humanity and its descendants, and the careful stewardship of the Earth become paramount articles of faith and morality.
- Patronise entertainment that presents a more creative, co-operative and less destructive narrative about humanity and how we solve our problems.
- Encourage your children to play games that create and restore, rather than kill or destroy; understand that play is but rehearsal for life and how children play may ultimately govern their fate.

10

The Getter of Wisdom (*Homo sapientior*)

To know yourself is the beginning of all wisdom

—Aristotle.

He who knows others is wise; he who knows himself is enlightened

—Lao Tzu.

The twenty-first century will present humanity with the supreme test of our wisdom. To attain it, to survive and prosper peaceably, we need to understand ourselves and our situation better.

Among the chief attributes of wisdom is the ability to interpret the likely future and to take precautions against adverse outcomes or take advantage of change and emerging opportunities. This ability to foresee events, based on past experience and a careful reading of present indicators, is the distinguishing quality of humans and has been the principal reason for our success so far. Now, however, we inhabit a time when that success has led to our predominating over the planet, in the process profoundly altering the very systems we rely on for survival: atmosphere, soil, water, other living creatures, our good health, our numbers and ability to get along with one another. We live in times when a domino-like succession of changes is occurring with such swiftness and hammer-blow ferocity that many people are dismayed, even paralysed by it. Nevertheless, we must all respond if we value our survival. This chapter explores two developments which will be essential to the continuance of both our civilisation and of humankind. It then offers a pathway forward.

© Springer International Publishing Switzerland 2017
J. Cribb, *Surviving the 21st Century*, DOI 10.1007/978-3-319-41270-2_10

Thinking as a Species

The existential threats of ecosystem collapse, resource depletion, nuclear war, climate change, global toxification, famine, overpopulation, pandemic disease, universal surveillance and uncontrollable technologies which surround us may find their solution in the most import human achievement of the past million years: the linking of minds, values, information and beliefs at lightspeed and in real time, around the planet.

This is a development without precedent, not only in our own history but also among all the species which have ever inhabited the Earth. Almost unawares, we are giving birth to an entirely different kind of human.

In the second trimester of a baby's gestation a marvellous thing happens. The neurons, axons and glia in the embryonic brain begin to connect—and cognition is born. An inanimate mass of cells and microbes becomes a sentient being, capable of thought, imagination, memory, logic, feelings, beliefs and dreams.

Today individual humans are connecting globally, at light speed, just like the cells in the embryonic brain. We are in the process of forming, if you like, a universal, Earth-sized mind. What the Jesuit philosopher and prehistorian, Teilhard de Chardin, once termed the 'noosphere'—the realm of human thought or intelligence that encircles the Earth—is becoming incarnate (de Chardin 1955). A higher understanding, and potentially a higher intellect, is in genesis—capable of interpreting, and maybe solving, our problems at *supra-human level* by applying millions of minds simultaneously to the issues, by sharing knowledge freely and by generating faster global understanding and consensus about what needs to be done. At the very moment in our social evolution when our national governments, corporations and existing institutions are seen to be failing to cope with the overwhelming challenges that encompass us, a new form of human interconnection and self-awareness has emerged that might, just possibly, save us from ourselves.

A million years ago, as we sat around the campfire on the African savannah, humans ensured our own survival by forging a complex society capable of identifying and overcoming the many threats that then surrounded us. Ten thousand years ago we began the process of conquering the threat of hunger through agriculture. This experience has carried through our entire existence, and is the mainspring of the creature we have become. We are very, very good at identifying potential risks and dangers, and finding collective ways to mitigate them. It's why we build hospitals and schools, have police, soldiers, firefighters and food inspectors, cleanse our water, process our waste, investigate air crashes, obey traffic lights. You could argue that threat avoidance in the name of survival is our greatest single attribute as a species.

Today the human species has never been more at risk, the result of our burgeoning population and the overgrowth in its demands on the Earth's natural resources and systems, combined with our natural aggression and competitiveness. Only a fool would imagine we can keep on behaving as we do today when there are ten billion of us, without grave risk to our entire civilisation and maybe our species.

Yet solutions to all of these challenges exist or can be developed.

The solution to population growth is already being implemented by the young women of the world, who are declining to marry and have babies in *all* societies. They are ignoring what men tell them. They are ignoring the priests, the patriarchs, the journalists, the politicians, the government baby-bribes. They are acting spontaneously to reduce their own fertility: in the 1950s, the average woman worldwide had 4.97 babies; by the 2010s this had halved to 2.4 (United Nations Population Division 2012). Something very great and very beautiful is happening among women, at the species level. Given worldwide support and approbation, and continued provision of education, healthcare and opportunity, women can lead the human population back to a sustainable level—between two (Daily et al. 1994) and four billion people— by the early twenty-second century *voluntarily*.

The solution to resource scarcity is recycling and re-use. As the population falls, there will be no more need of mining—all the metals, nutrients and materials we will ever need will be available and readily accessible in our waste stream. We just need to 'mine' that, remove the toxins and recycle materials and water endlessly.

The solution to the resource problems created by our money-driven system is to dematerialise wealth—to build an economy founded on products of the mind, rather than on material goods. To employ people in creative industries, rather than in anachronistic labour like manufacturing, mining and physical agriculture (which will mostly be performed robotically). This way, money will not be used to destroy finite things of real value such as soil, water, biodiversity and the atmosphere, as it is presently being used to do. Money, being immaterial and infinite, can be used to create products and services that are equally immaterial and infinite—products of the human imagination, which are the true future wealth of society.

The solution to the current wave of extinction, and to food security, is to re-wild half our current farmed area, and develop clean, intensive food systems in our cities. Then we must pay farmers and indigenous people to look after the wildlife, the soil carbon, the vegetation, the small water cycle, the genetic diversity and all the other eco-services we depend on for our survival, to be the Stewards of the Earth. They are the ones best qualified to do this.

The solution to both climate change and to the pandemic poisoning of all humans and life on Earth is the same—we can achieve both by eliminating use of oil, gas and coal, and embracing renewable energy and algae technology for fuel, food, fibre, industrial chemicals, plastics and medications. We can cleanse the world together by demanding products that are safe and healthy, and rewarding for the companies and farmers who produce them.

The obstacle we face is that few national governments in the world are likely to embrace such a program wholeheartedly. They are dinosaurs, bogged in a tar-pit[1] of empty opinion polls, imprisoned in an echo chamber of distorted information and often beholden to myopic vested interests and selfish economic powers which resist change.

So how do we solve these mega challenges?

In 2017 there are 3.6 billion internet users on Earth and by the 2040s most people will be online (Hettick 2013). Young people are reaching out to one another in real time, across the divides of race, nationality, ethnicity, religious belief, language and prejudice. They are learning how alike we all are. How many things we share. How we can 'like', help, support and depend on each other. They are also learning how deadly are the prejudices, the ignorance, the fears and the hatreds of their parents towards other humans. And how pointless.

The antidotes to ignorance, fear and hate are knowledge, understanding and familiarity. The internet, in spite of its shortcomings, can supply both. Humanity is still in the second trimester of the formation of a universal mind—a connected humanity capable of collective thought, information sharing and resolute collaborative action. It is often argued by scientists that social media is full of rubbish, trivia, abuse and misinformation—but so too is the average human brain. Most of us make our way through life, as individuals, by sorting the sensible, useful, ethical stuff from the rubbish—by choosing the generous over the petty, the altruistic over the selfish, the practical over the delusional. If we can do it within our own minds, we can do it with an Earth-sized mind of which our own brains are but individual cells, networked with billions of others.

We stand at the threshold of our pan-species era. It is far too early to dismiss such a development, as some might be tempted to do if they were unaware of the power and reach of social media and the web. Be sceptical, by all means, but also be open to the possibilities.

---

[1] I am indebted to former US Army general Norman Schwarzkopf for this colourful, if palaeontologically inexact, expression which he first applied in a TV interview to the consequences for the US of invading Iraq and capturing Baghdad. It was too apt not to re-use.

Dream the dream of humanity starting to think together as a species.

To creatures accustomed to regarding ourselves as individuals, the idea of being part of a greater organism may seem eerie, even threatening. Yet biologists have known for some time that we 'individuals' are in fact assemblages of separate cells, DNA and microbes which coalesce for a time to form a person, an animal or a plant. There are ancient viral genes interlaced with our own genome (Wildshutte et al. 2016) and colonies of bacteria in our gut working to preserve our health, as well as the different types of cells in our body. Each 'individual' is effectively, an ecosystem or a biological corporation (Clark 2012). It should not therefore be too hard to regard ourselves as part of a larger organism whose survival depends on mutual co-operation and shared wisdom.

As things stand we are still in the kindergarten phase of learning the art of common thought, of developing the universal reasoning capable of understanding and solving our mutual challenges. Social media, which critics often dismiss as trivial and of no account, are taking over from traditional forms of controlled media and politics in extraordinary ways. They are a grand reflection of our common values, as well as our vices, pettiness and shortcomings. They are now used by governments, aid agencies, charities and the United Nations as well as rock stars and celebrity air-heads.

Out of this inchoate planetary chatter, common threads of thought are already emerging. Through the internet, knowledge once held only by elites is being exchanged, values shared, attitudes reshaped—and from this, gradually, a worldwide consensus for action is forming. In science, for example, massively more scientific publications are now written for and shared with the general public via social media instead of being stashed away in musty journals in the mediaeval libraries that still clog our university campuses— and exclude 'the rabble'. The light of science is falling on Earth's citizens wherever the internet reaches. For all the internet trolls and sociopaths, there are also tens of millions of decent citizens and intelligent, caring individuals sharing their moral values and goodwill at lightspeed: those values will prevail.

If a consensus of a majority of humans starts to emerge on *any* of the major existential threats that confront us, it will be an act like no other in history.

It will be one that no government, no corporation, institution or society can ignore.

It will be more powerful than nations or governments—because they will have no power over it.

Its influence will exceed that of the great religions or political movements. It will be economically more potent than the largest multinational corporations.

As we are already seeing in issues such as industrial slavery,[2] fair trade and ethical consumerism (World Fair Trade Organization 2016), the views and values of millions of concerned consumers can change the way industry behaves, the products and services it produces and the rules by which it operates. They can make unethical, dirty, cruel, unsustainable and harmful practices a bad business decision for the people who practice them. This will apply market forces in unprecedented ways to cleanse our energy system, our food supply and our toxic planet. It will also compel corporates to reward their CEOs according to their performance in ways that a wider humanity approves, rather than mining the planet and their own company for short-term profits.

It will do this by exerting a force that many conservatives, including academics and scientists, despise: fashion. Fashion is not just about smart clothes and trendy adornments. It is also about ideas and values expressed in our consumer choices and the lives we choose to lead, the technologies we prefer, our political opinions. It attracts mass attention because it is the bow-wave of change, innovation, creativity and fresh public opinion. It can be about serious and important changes, as well as trivial and flippant ones. If having fewer babies, seeking cleaner food, safer, more ethical products and rejecting fossil fuels become universal fashions among the young, and are followed by billions of people engaged in shaping a safer future, it will change how global society is regulated, how the world economy functions and the signals it sends to both corporations and governments worldwide.

Such a consensus will be more influential on the human destiny than any power or principality heretofore. Such a discourse will give new life to the collaborative rather than the competitive aspect of our nature.

Can we make it work? The simple answer is that, if we don't, then fewer than a billion humans will probably inhabit the famine-, disease-, climate- and war-ravaged ruins of our planet a 100 years from now (Schellnhuber 2009). We have the strongest of all possible motives to succeed. The one all of our ancestors would have best understood. Survival.

This is the time when we get to choose whether we are truly *Homo sapiens sapiens*—or some other organism, that failed the Darwinian test.

---

[2] see, for example, Modern Slavery, https://modernslavery.co.uk/index.html.

The Age of Women

If women led the world, it would probably be vastly less toxic, far less prone to climate change, hunger, war and environmental devastation. Far less at risk from its own 'success'. This reflection was drawn from the research for an earlier book, *Poisoned Planet* (Cribb 2014) in which it became painfully evident that the 250 billion tonnes of chemical substances emitted annually by human activity—arguably our greatest impact of all upon the planet (Chap. 6)—are almost exclusively the handiwork of men, as distinct from women. This is not to say that women don't benefit from these activities or even, often, approve them. But they rarely drive them, at least with such sanguine (and sanguinary) disregard for present and future generations.

Chemistry is a profession that has long experienced absolute male dominance. It began with fairly harmless products like dyes and textile treatments but rapidly progressed to thoroughly masculine things like high explosives, poison gas and the ingredients for atomic weapons. Now it has moved on to the mass production of hormone disruptors, cancer-causing agents and nerve poisons on such a universal scale that, without radical reform, they will probably affect every child on the planet for the rest of history. Today, while it is obligatory to test a new aircraft, car or mobile phone for safety, the majority of our 144,000 chemicals have never been fully tested, or in most cases tested at all, according to the United Nations Environment Program (UNEP). Yet they are released into our living environment, and us, anyway.

Globally only four out of 166 Nobel laureates for chemistry in the past century or so were women, and in the US, a leader in equal opportunity, women made up only 16 % of tenured chemistry academics and only 9 % of chemical company CEOs. It is likely that the gender imbalance in places such as Japan, Korea, India and China (the world's poorly-regulated chemical powerhouses of coming decades) is even greater. And although more young women are studying chemistry at university and occupy the lower rungs of the profession than in the past, they are not responsible for the big decisions about what gets put into the living environment or the human species, and whether or not it has been fully tested for safety—especially for children.

On the other hand, if you take a quick look around social media and the cybersphere you will find that most of the organisations of parents, citizens, consumers and victims which are most concerned with health, with children's wellbeing and with rolling back the tide of toxic contamination in our lives are led by women. When it comes to assessing the risks and rewards of chemistry, male and female thinking clearly diverge.

Climate change, too, is a gender issue—as much as one of physics or economics. The vast majority of people who release carbon for a living, or who cause it to be released, or who then burn it excessively, are males— miners, foresters, big farmers, builders, pilots, racing car drivers. *Men are driving global warming, not women.* In Australia, for example, women comprise only 5% of the top five mining professions (Minerals Council of Australia 2013). And most countries worldwide would have far fewer females in those professions. Like chemistry, mining remains an industry utterly dominated by masculine thought processes.

Likewise, the vast majority of those who clear-fell forests or degrade topsoil, contaminate rivers and air, fish out the oceans and exterminate wildlife are men, not women. Though it is claimed that over half of the world's 1.4 billion farmers are women, most of those who operate the big soil-destroying equipment and spray the most chemicals, or who serve as farm leaders, are men.

History demonstrates that males prefer immediate, direct, vigorous mechanical or chemical action to solve a problem and obtain a short-term goal—whether it is defeating an enemy, growing a crop or a nation, building a skyscraper or a trading enterprise. Men have a history as risk-takers. From the time we emerged on the African savannah males have been conditioned to sacrifice themselves, along with anyone who gets in their road, for perceived short-term advantage—a training horrifically attested in two world wars. Men tend to accentuate the competitive side of the human personality.

Female thinking usually looks further and wider, to the impact of our choices on children, grandchildren, society, food, water, health, living conditions and the environment. As a rule, females tend not to start wars (though they may support them). They prefer co-operative solutions.

Thus, certain men are happy to mine coal for instant prosperity—and gamble with the future of their own and everyone else's grandkids. Male thought tends to value immediate over future consequences, cash and prosperity now rather than a safe, stable, healthy life later. Of course, this is not a hard and fast rule, nor a genuine stereotype—there are countless women and men who embrace the alternate mode of thought or are positioned all the way along the spectrum. But ask yourself this: if the vast majority of the world's miners, chemists, foresters, fishers, big farmers, builders, bankers or soldiers were women, how would *they* run these vital industries? How would *they* assess the balance of risk and reward?

Nor is this to decry males or the way they think. For generations pragmatic masculine solutions—machinery, weapons, chemicals, drugs, transport, energy supplies, agriculture, dams—have provided swift, practical answers to the immediate needs and wants of society. Men were the

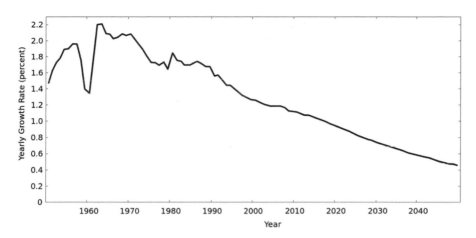

Fig. 10.1 The rate of growth in the human population: women have already made the big decision. *Source:* Wikimedia Commons

chief architects of the world's religions, political, monetary, food, military and social systems. This masculine predomination worked fine for as long as human numbers were small and our demands on the planet's life support systems relatively modest.

Yet, in the space of a single generation, *everything* has changed.

Today male-dominated production and exploitation is playing havoc with the composition of the Earth's atmosphere and oceans, with human health, with the survival of other species and whole landscapes. As the population surges towards 10 or 11 billion and the global economy expands, these impacts are likely to *more than double* over the course of the century.

Interestingly, women worldwide have already taken a decision to ignore men completely in the one area over which they have real control—their own fertility (Fig. 10.1). Female fertility—while still high in some places—is dropping universally in spite of all the patriarchal lectures and juicy bribes offered by male-dominated governments. The UN Population Division says the number of children per woman has fallen from 4.97 in the 1950s to 3.45 in the 1980s, to 2.6 in the 2000s, 2.45 in the 2010s, and will continue to decline to a forecast 2.2 by the 2050s and 1.9 in the 2090s (United Nations Population Division 2012). Women have, apparently, already taken the species-level 'decision' that a safer future involves fewer people; though it may take another century or so to bring about the necessary reduction to a sustainable population. (Ironically, male fertility is dropping too—probably as a result of the toxic avalanche, which is chemically feminizing males around the world, lowering sperm counts and is thought by some scientists to be a factor in changing sexual preferences, which also tend to lower reproduction (Carpenter et al. 2002b). But not fast enough to make a difference.)

So great is the human impact on the biosphere going to be by the second part of the twenty-first century that traditional male-think—essential in earlier times for survival and growth—becomes a disadvantage, and female-think (by men as well as women) a major dynamic in the prospects for the continuance and wellbeing of a civilisation that needs to share, recycle, sustain, heal, co-operate and mutually understand more than it needs to over-produce, compete and exploit. All the more reason for concern, then, that women are still excluded from power and policy by so many societies, corporations, religions and organisations. These are denying themselves the very thought processes most needed for survival under the altered circumstances of the twenty-first century.

This isn't an argument for equal opportunity, or even feminism. It isn't about politics. It's more important than those. It's about the *emerging rules for human survival* and wellbeing in a finite and increasingly damaged world. It's about something we all need to take on board, regardless of our gender.

Men have been the leaders in most of the past great achievements of civilisation: the stone, bronze, iron, agricultural, industrial and IT revolutions were substantially masculine artefacts. In their day they were essential to get us to where we are now, a place both wonderful—and filled with growing perils.

To secure our future we need a new stamp of leadership—one which takes a longer-term view, mitigates risks, protects, sustains, collaborates and cares more thoughtfully for posterity and the planet on which it will depend. Leadership in the twenty-first century—the era of 'peak everything'—will demand different skills, mindsets and qualities to that of the 20th. We must all embody them if we are to lead humankind out of the place of peril to the place of hope, prosperity and safety. Above all, this depends upon the leadership of young women.

To survive, humanity now needs The Age of Women.

The Earth Charter

One essential element in human survival in the C21st is the need to share a common vision of the way ahead, to unite and inspire us. There have been numerous attempts by many authors to articulate this, and among the best is The Earth Charter which is "an ethical framework for building a just, sustainable, and peaceful global society in the 21st century. It seeks to inspire in all people a new sense of global interdependence and shared responsibility for the well-being of the whole human family, the greater community of life, and future generations. It is a vision of hope and a call to action."

Springing out of the 1987 Brundtland Report *"Our Common Future"* the Charter began as a United Nations initiative pushed by leaders such as Maurice Strong and Mikhail Gorbachev, and was carried forward and completed as a global civil society initiative. It was launched on 29 June, 2000, in a ceremony at the Peace Palace, in The Hague. It is generally regarded as a global consensus statement about sustainability, equity and sustainable development.

The Charter has 16 principles, starting with the need to respect the Earth and all in it, built just democratic societies for all, protect the Earth's resources, life and beauty for the future, avoid overconsumption and destruction of resources, eliminate poverty and gender inequality, support human health and dignity, educate all and promote peace. The full charter is available here: http://earthcharter.org/invent/images/uploads/echarter_english.pdf

The Earth Charter is, in essence, an appeal to human wisdom. However, like many aspirational documents, it risks being ignored by the bulk of humanity as we go about our daily lives, without a clear-eyed appreciation of the scale, number and compounding impact of the existential perils we now face, which this book has described, or of the need for cross-cutting solutions that address them all, not just a few of them. Both the vision and an informed understanding of our situation by all humans are essential steps on the path to wisdom.

The Sustainable Development Goals

The 17 Millennium Development Goals of the United Nations, declared in 2000 and then updated in 2015 as the Sustainable Development Goals (SDGs), are a call to action for peaceful, sustainable development to which most countries have signed on. They are formed around the four basic elements: people, planet, prosperity and peace.[3]

The Goals are:

1. End poverty
2. End hunger and promote sustainable agriculture
3. Ensure healthy lives
4. Inclusive education
5. Greater gender equality
6. Sustainable water supplies
7. Sustainable energy supplies

---

[3] For full details, see http://www.un.org/ga/search/view_doc.asp?symbol=A/69/L.85&Lang=E.

8. Sustainable economic growth
9. Resilient infrastructure
10. Reduced inequality
11. Sustainable cities
12. Sustainable consumption
13. Combat climate change
14. Conserve the oceans
15. Conserve land-based ecosystems
16. Promote peaceful, just, inclusive societies
17. Form global partnerships.

Though the SDGs acknowledge that "The survival of many societies, and of the biological support systems of the planet, is at risk", they fail to spell out that, thanks to the constellation of existential threats, civilization—and maybe even the human species itself—is in jeopardy. While many of the people who contributed to them know this full well, the tendency of global institutions to employ soothing, negotiated, diplomatic language in preference to the plain truth, leaves an impression that the situation is less than critical, that there is ample time to make incremental improvements in all these aspirational goals. This soft-pedalling allows many people, societies and nations to ignore or downplay the clear and present dangers documented in this book in favour of their day-to-day, and generally local, concerns.

In particular, the SDGs fail to explicitly address the omnipresent threat of extinction from a renewed WMD arms race, the risk of global warming going into planetary overdrive, the threat posed by the universal poisoning of humanity, its children and global biota, the need to reduce human numbers, dematerialise the economy, shift food production from farms to cities, recycle everything, regulate dangerous science and overcome the crippling effect of delusional beliefs. They also suffer from the political adversity that they are at odds with the selfish interests of many nations and corporations, who may pay them lip service, but will as a rule do little to implement them—if they do not actively seek to undermine them.

They are essential steps on the path to wisdom, but cover only part of the journey.

The following roadmaps (Figs. 10.2 and 10.3), built from the advice offered in earlier chapters, indicate the kinds of actions at both species and personal level which will help to enhance humanity's prospects for survival and prosperity in the twenty-first century.

These roadmaps indicate what each of us can do to improve our own and our species' chances of survival in the face of growing existential risks.

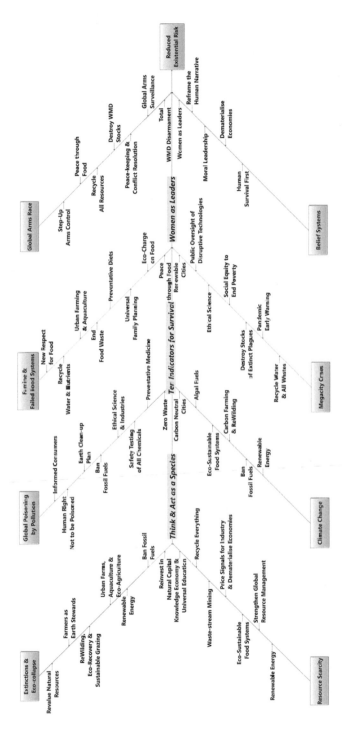

Fig. 10.2 Roadmap for human survival in the twenty-first century. Credit: Peter Day, 2016. For explanation, see Glossary for Roadmaps, p. ….

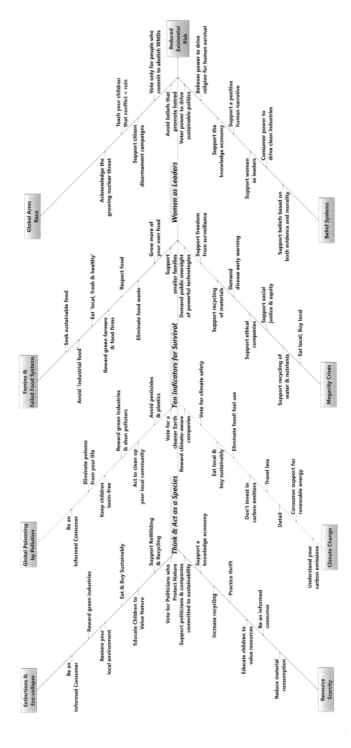

Fig. 10.3 Personal roadmap to assist in ensuring human survival, health and prosperity in the twenty-first century. Credit: Peter Day, 2016. For explanation, see Glossary for Roadmaps, p. …

Measuring Wisdom

There is a limit to how many people the Earth can carry because there are limits to the resources and systems it has to sustain them. Any thinking person knows this. What is largely unknown is exactly where those limits, or boundaries, lie—although we are now getting a much clearer, and more deeply disturbing, picture of the consequences of breaching them from a host of scientific measurements.

That we should establish the boundaries for various critical factors on which life depends was proposed by Rockstrom, Steffen, Schellnhuber, Hughes and their colleagues in two groundbreaking papers, referred to in Chaps. 2 and 9 (Steffen et al. 2015; Rockström et al. 2009). In the first, they stated:

> Anthropogenic pressures on the Earth System have reached a scale where abrupt global environmental change can no longer be excluded. We propose a new approach to global sustainability in which we define planetary boundaries within which we expect that humanity can operate safely. Transgressing one or more planetary boundaries may be deleterious or even catastrophic due to the risk of crossing thresholds that will trigger non-linear, abrupt environmental change within continental- to planetary-scale systems.

They proposed the setting of nine environmental boundaries beyond which humanity ought not to venture, for its own safety:

- climate change (measured by CO_2 concentration in the atmosphere)
- ocean acidification (measured by the pH of seawater)
- stratospheric ozone levels
- levels of nitrogen and phosphorus cycling in the Earth system
- global freshwater use (<4000 km^3 $year^{-1}$ of consumptive use of runoff resources)
- land system changes (ratio of land used for farming and cities)
- loss of species worldwide
- 'novel entities' including new chemicals, GMOs and other products of synthetic biology and artificial intelligence
- levels of global air pollution.

The team noted we had already crossed two of these boundaries: species extinction and nitrogen emissions and were approaching two more, in climate and land-use.

These boundaries (Fig. 10.4) are, in effect, a report card on how we humans are managing the Earth and its life support systems. They provide us with

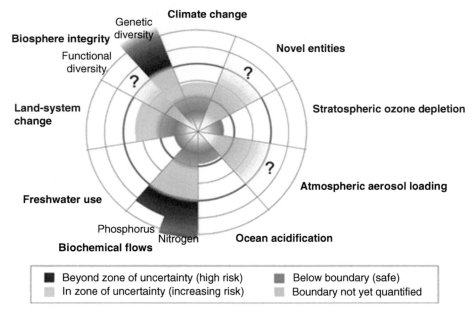

Fig. 10.4 The global boundaries proposed by Rockstrom et al., which humanity for the sake of its own survival ought not to breach. *Source:* Nature 2009

clear and unequivocal warnings, based on verifiable measurements, of when we are approaching a tipping point beyond which it is not safe for us or the Earth to venture. They offer no prescriptions for *what* we should do but, like a traffic light, simply advise us *when* to slow down, when to stop if we wish to avoid a smash, and when to back up.

Equally, as a report card, they offer us a splendid way to monitor our progress in setting things to rights. They are a first step towards creating a common language of caring for the Earth and our own posterity.

Unwise Humans

In the opening chapter, we looked at how humans came to name themselves 'wise'. It was a thin argument even in the 1750s, and in the light of all we are doing to the planet and ourselves, thinner still today. If the previous eight chapters have done nothing else they have at least brought together some of the evidence that, as a species, we are not very wise. Individuals, maybe. A few. But not the human species.

This book contends that *Homo* does not merit the descriptors *sapiens sapiens*, on strict scientific or any other grounds. As noted in Chap. 1, there are plenty of precedents for changing the name of a species—especially if the name turns out to be totally inappropriate or worse, dangerously misleading.

This book proposes that the human species be re-named, and our self-admiring title *Homo sapiens sapiens* be abolished, stricken from the biological canon. Recognising our nature and its consequences, we should have both the wisdom and the honesty to demote ourselves.

It is entirely fitting, in this age of global communication, that the new name for our kind should be the subject of a worldwide discussion and debate. Indeed, that is the very purpose of the suggestion: to bring as many people as possible from all cultures, walks of life and nationalities into the conversation, to consider how our collective actions describe us. To debate our true nature. It should not be a matter left to a handful of scholars in a dark corner. This is not a scientific nicety, nor one of adherence to the rulebook of the International Code on Zoological Nomenclature (ICZN). It is a matter of life or death for billions. A humanity which collectively recognises its errors, its unwisdom and the perils our behaviours are bringing upon us is much more likely to survive and prosper than one which doesn't.

Trading in an antique Latin name on something more modern might seem of scant significance to the vast majority of people, it can be fairly argued. But then again it might just change the way we see ourselves. A name is an essential part of who we think we are. If your much-loved grandparent declared that you were no longer fit to bear the family name, and cast you out because of your misdeeds, unless you had the hide of a rhino you might at least reflect on it. You might at least ask how you can gain re-admittance to the fold. Losing one's name is a very special form of shaming, because with it you lose a part of your identity, your self-image, your pride, your associations, how you understand yourself—and that is a serious matter.

The piece of our identity which humans most need to lose is the cocky, conceited, thoughtless, "we're smart, we got all the answers" attitude. This time, the mounting weight of scientific evidence suggests, we don't have all the answers—only a few, and those depend on a degree of mutual co-operation and consensus the world has never before seen. As a *species* we are not wise. We are not smart. We may not even be intelligent enough to ensure our own long-term existence. That remains to be seen.

So let's look ourselves straight in the eye, tell ourselves the truth for a change—and drop the *sapiens*. It would be something that almost everyone in the world would talk about, and many would think hard about.

Losing our name need not, however, be a permanent demotion.

The option will exist for humanity to earn back our *sapiens* by proving that we deserve it, that we are, in fact, quite wise. And the way to do that is to limit our impact on the planet so that it falls within 'safe boundaries' like those proposed by Johan Rockstom and colleagues. In the light of the somewhat wider evidence presented in this book about threats to the human future, ten criteria are here proposed by which we may judge ourselves.

Ten Ways to Measure Our Survival Prospects in the C21st

1. progress in banning and eliminating nuclear and all weapons of mass destruction, their waste and precursor substances
2. rate of removal of CO_2, air toxics, ozone depleters and particulates from the atmosphere
3. rate of protection, cleansing and recycling of the world's fresh water, nutrients and materials
4. progress in the clean-up of our chemical environment, including elimination of all known cancer-causing and major toxic substances, fossil fuels, lost nutrients and mineral wastes
5. the ending of the 6th Extinction through the progressive re-wilding of half of the world's presently cleared landscapes under a global stewardship plan employing farmers and indigenous peoples
6. rate of conversion of the world's food supply to urban agriculture, soilless farming and aquaculture; increases in healthy, fresh, diverse, local food output
7. the assumption by women of leadership roles in government, industry, religion and all walks of society
8. rate of reduction in human numbers to a sustainable population
9. the rate of recycling and regeneration of the Earth's limited resources of nutrients, water, minerals, energy, forests and wildlife
10. open, ethical control by society over disruptive technologies including synthetic biology, artificial intelligence, armaments and global surveillance of individuals.

By measuring our progress towards each of these goals we may not only secure the longer-term survival of both civilisation and humans, but also that of the planet in the safe, stable, well-supplied and temperate condition in which it gave birth to us. In each case, the setting of a clear target lets us know how far we have to go to achieve it—and offers a common incentive to try harder.

So the second principal proposal of this book is the establishment of firm, measurable criteria which protect humanity, the Earth, its many systems which support us, and the marvels of life which it contains. Which set clear boundaries which we dare not exceed, for the sake of all our children yet unborn. Which reward us for reducing our negative impact on them and on the planet.

Under this proposition every government, every family, and indeed every citizen will receive a regular update on how well or poorly we are doing. The ten survival indicators—and how well or badly we are doing in achieving them—should be on every smartphone, on the nightly broadcast news, in the mouths of every religious leader and politician, on the sides of milk cartons and food packaging, in our school books and as sustainability indicators for every product, like the energy stars on a fridge or the fuel efficiency rating of a car. They should be followed more eagerly than the bulls and bears of the stock market, the gyrations of the money market, the fortunes of our favourite sports teams and the daily weather report. These simple indicators, and the message they bear about our chances of survival, should be everywhere, unavoidable, keyed to all our main activities, especially our actions as consumers and investors.

In this way each of us will become more conscious of our responsibility for our planet and our species, as well as for ourselves. More engaged in the task of charting our own future. More conscious of the cost of our actions and decisions to the planet and to posterity. More motivated to try harder and do better as individuals. With luck, the ten indicators will play to both humanity's competitive and collaborative instincts by setting worthy, safe, wise goals for us to achieve together, giving us grounds for global celebration as we gain them through our common efforts. They can be the basis for countries, creeds, corporations and communities to compete peacefully with one another to excel, show leadership, inspire, achieve, cleanse, heal, collaborate and protect.

Foresight is humanity's ultimate skill. Our quintessential wisdom is the wisdom of the survivor.

The question the twenty-first century will answer is: do humans have it still?

Glossary for Roadmaps

This glossary provides further explanation of terms used in the Roadmaps, Figs. 10.2 and 10.3.

Algal oils—renewable oil extracted from microscopic water plants (algae) which can replace petroleum products.

Aquaculture—farming of fish and water plants on land and at sea, using natural feeds derived from farmed algae.

Ban fossil fuels—complete cessation of all fossil fuel use (for energy and products such as plastics or drugs) by 2030 and their replacement with safer renewables. All government subsidies to the fossil fuels industry to cease.

Clean Up The Earth—worldwide collaboration by parents, consumers, industries and governments to eradicate toxins from all products, foods and emissions and reduce toxic exposure of our children.

Dematerialisation—progressively convert the world and national economies to run on products of the human mind, rather than the overuse and destruction of material resources such as air, water, soil, forests, fish, wildlife.

Destroy plagues—eliminate all stocks of extinct disease pathogens held in military laboratories; outlaw the scientific development of new plagues.

Nuclear disarmament—outlaw and destroy all nuclear, chemical and biological weapons and stocks of materials. Convert uranium nuclear energy to renewables or safe nuclear (eg. fusion).

Disease prevention—refocus medicine and science around disease prevention via diet, healthy environments, exercise, eco-health and exclusion of toxins etc (instead of chemical 'cures').

Eco-agriculture—the contraction of agriculture to the best farming areas where its operates sustainably, emphasising soil carbon retention, surface cover, water and nutrient recycling and minimal use of biocides, and where farmers are appropriately rewarded.

Eco-recovery—international collaboration to restore great forests, grasslands, rangelands, fresh waters, oceans, soils and ecosystems. Also see: Clean Up The Earth.

Education—establish universal, free, education to ensure each of Earth's citizens understands the need to sustain and care for vital Earth systems and resources that support them, and send the right signals to industry and politics.

End food waste—recycle all nutrients, especially in cities.

Ethical industries—use market signals to encourage industries to adopt clean, safe, ethical products and processes, recycle materials and respect the right of future generations to a sustainable world. Reward industries which adopt approaches such as green chemistry, product stewardship and zero waste by preferring their products.

Ethical research: require all young scientists and engineers to take a profession oath to 'first, do no harm'. Educate them in their ethical responsibilities.

Family planning—ensure the availability of family planning, education and healthcare for women in all societies.

Green cities—redesign the world's cities so they recycle 100% of their water, nutrients, metals and building materials, grow <50% of their own food, are carbon-neutral and green with vegetation and wildlife.

Global arms watch—develop stronger, more collaborative global surveillance of nations and groups who pose a potential risk of WMD terrorism. Develop a global citizens' peace movement operating in all countries and societies to warn of the dangers of continued retention of WMD and exert political pressure for their abolition.

Global biosecurity—develop a global surveillance network to combat the spread of pests, weeds and pandemic diseases.

Healthy diet—reshape the world diet from one that degrades the planet and our personal health to one which protects and preserves both.

Human Rights: introduce a universal human right not to be poisoned. Establish a new human right prohibiting the mass surveillance of entire populations and restricting cradle-to-grave data collection on individuals not suspected of a crime.

Human survival first—refocus world, national and local politics on human survival through citizen pressure.

Knowledge economy—reshape the world and national economies so they depend less on physical resources and more on growth in products of the inexhaustible human intellect..

Knowledge sharing—establish a universal internet-based system for sharing information and knowledge about food and material products so consumers can make wise, healthy and sustainable choices which send the appropriate incentives to industry and signals to government.

Control supertechnologies—public oversight of all developments in artificial intelligence (AI), killer robots, nanotechnology, quantum computing and biotechnology. Outlaw the scientific development of novel pathogens and technologies with potential to harm humans. Impose a code of ethics and public transparency on all scientific research.

Female leadership—adopt female principles in the leadership of all major industries, governments, corporations and religions. Promote women to positions of global leadership.

Moral leadership—encourage the world's religions to take more courageous moral leadership over issues that threaten humanity as a whole, and to set aside their differences for the common weal.

Pandemic early warning—establish a worldwide early warning systems for new pandemics. Publicly fund a global endeavour to develop new antibiotics and antivirals.

Peace-keeping—develop stronger national and international rules and institutions for conflict resolution. End poverty as a precursor to a more stable, cohesive and peaceful world. Reinvest military budgets in food security.

Peace through food—assign a fixed percentage of military budgets to peacekeeping through the preservation of the food supply. Reinvigorate food and agricultural research worldwide, especially to accelerate urban farming, soilless farming and aquaculture.

Recycling—recycle everything: water, nutrients, metals, timber, plastics, textiles and building materials etc. Turn all food system waste back into nutrients for food production.

Sustainable Food Systems—renewable food production from eco-agriculture, urban agriculture and aquaculture, which recycles water and nutrients, stores carbon and end soil damage.

Human Narrative—reframe our economic, political, religious and entertainment narratives to make the survival of civilisation and humanity the primary goal. Inspire a new narrative about humans which prizes co-operation, tolerance, restoration, cleansing, saving and sustaining over divisive, selfish and destructive values.

Renewable energy—replace fossil fuels with electricity from renewable sources (eg solar, wind, tide, geothermal, stored in clean, efficient batteries). Accelerate worldwide research and investment in clean, renewable energy. Replace fossil transport fuels with algal oil. Replace uranium-cycle with thorium-cycle nuclear.

Rewilding—Cease clearance of forests and grasslands. Return up to half the Earth's cleared lands to wilderness, native flora and fauna managed by indigenous people and farmers.

Urban farms—intensive, sustainable food production systems, using minimal land and pesticides and based on recycled water and nutrients, including food waste. 'Agritecture'—combining cities with food production and vertical farms

Value natural resources—prevent ecological collapse and repair landscapes through a small levy on all food and consumables. This will cover rewilding and resource care by farmers and indigenous people. Raise the next generation of humans to respect food. Restructure economics of the global food chain to promote sustainable, clean production and educated consumers.

Zero waste—ban the permanent dumping of waste and replace it with waste-stream mining.

Zero toxics—eliminate use of all known toxics from the food chain, water supplies, personal care products, home goods and the wider environment. Mandate toxicity testing of all new industrial substances, mixtures and emissions.

References

Aleksandrov VV, Stenchikov GL (1983) On the modelling of the climatic consequences of the nuclear war. In: Proceedings of applied mathematics, computing centre, USSR Academy of Sciences, Moscow, 21 pp

Allen JG et al (2015) Associations of cognitive function scores with carbon dioxide, ventilation, and volatile organic compound exposures in office workers: a controlled exposure study of Green and Conventional Office environments. Environ Health Perspect. doi:10.1289/ehp.1510037. http://ehp.niehs.nih.gov/15-10037/

Alston J, Beddow JM, Pardey P (2009) Mendel versus Malthus: research, productivity and food prices in the long run. University of Minnesota. http://ageconsearch. umn.edu/bitstream/53400/2/SP-IP-09-01.pdf

American Nutrition Association (2010) USDA defines food deserts. Nutrition Digest. http://americannutritionassociation.org/newsletter/usda-defines-food-deserts

Anderson S, Cavanagh J (2000) Top 200: the rise of Global Corporate Power. Global Policy Forum. https://www.globalpolicy.org/component/content/article/221/47211. html

Andresen CG, Lougheed VL (2015) Disappearing Arctic tundra ponds: fine-scale analysis of surface hydrology in drained thaw lake basins over a 65 year period (1948–2013). J Geophys Res Biogeosci 120:466–479. doi:10.1002/2014JG002778

Annan K (2003) Foreword to Living with risk: a global review of disaster reduction initiatives. UN/ISDR 2004. http://www.unisdr.org/we/inform/publications/657

Anon. Descartes never saw an ape. Classes in anthropogenesis, or the difference between human and animal. University of Warsaw, http://informatorects.uw.edu. pl/en/courses/view?prz_kod=4018-KONW85-OG

Anthony S (2014) The solar storm of 2012 that almost sent us back to a postapocalyptic Stone Age. Extremetech. 24 July 2014. http://www.extremetech.com/ extreme/186805-the-solar-storm-of-2012-that-almost-sent-us-back-to-a-postapocalyptic-stone-age

Arms Control Association (2014a) Organisation for the prohibition of chemical weapons 2015. http://www.opcw.org/our-work/demilitarisation/destruction-of-chemical-weapons/

Arms Control Association (2014b) Chemical and biological weapons status at a glance. http://www.armscontrol.org/factsheets/cbwprolif

Arms Control Association (2015) Nuclear weapons: who has what at a glance. Arms Control Association fact sheet. https://www.armscontrol.org/factsheets/Nuclear-weaponswhohaswhat

Asseng S et al (2014) Rising temperatures reduce global wheat production. Nature Climate Change. doi:10.1038/NCLIMATE2470

Auman H et al (1997) PCBS, DDE, DDT, and TCDD—EQ in two species of albatross on Sand Island, Midway Atoll, North Pacific Ocean. Environ Toxicol Chem 16(3):498–504

Australian Antarctic Division (2012) Pollution and waste, August 2012, www.antarctica.gov.au/environment/pollution-and-waste

Australian Institute of Marine Science (2012) The Great Barrier Reef has lost half of its coral in the last 27 years. AIMS Press Release, 2 October 2012, http://www.aims.gov.au/docs/media/latest-releases/-/asset_publisher/8Kfw/content/2-october-2012-the-great-barrier-reef-has-lost-half-of-its-coral-in-the-last-27-years

Australian Rock Art Initiative (2011) The Bradshaw paintings. http://www.bradshawfoundation.com/bradshaws/bradshaw_paintings.php

Bai ZG, Dent DL, Olsson L, Schaepman ME (2008) Global assessment of land degradation and improvement 1: identification by remote sensing. Report 2008/01, FAO/ISRIC Rome/Wageningen

Barbut M (2014) Land degradation—a security issue. UNCCD News 6, p2. http://newsbox.unccd.int/imgissue/UNCCDNews6_1.pdf

BBC (2015) Not in front of the telly: warning over 'listening' TV, 9 February 2015. http://www.bbc.com/news/technology-31296188

Beattie A (2015) The history of money: from Barter to Banknotes. Investopedia. http://www.investopedia.com/articles/07/roots_of_money.asp

Benedictow OJ (2005) The Black Death: the greatest catastrophe ever. History Today 55(3). http://www.historytoday.com/ole-j-benedictow/black-death-greatest-catastrophe-ever

Berne S, Marchand M, D'Ozouville L (1980) Pollution of sea water and marine sediments in coastal areas. Ambio 9(6):287–293, Retrieved from http://www.jstor.org/stable/4312607

Biello D (2012) Farmers deplete fossil water in world's breadbaskets. Scientific American. 9 August 2012. http://blogs.scientificamerican.com/observations/2012/08/09/farmers-deplete-fossil-water-in-worlds-breadbaskets/

Blair D (2015) North Korea's nuclear arsenal bigger than feared. UK Daily Telegraph. http://www.canberratimes.com.au/world/north-koreas-nuclear-arsenal-bigger-than-feared-20150424-1ms7fg.html. Accessed 24 Apr 2015

Bloomberg (2015) Quantum computers entice Wall Street vowing higher returns, 9 December2015.http://www.bloomberg.com/news/articles/2015-12-09/quantum-supercomputers-entice-wall-street-vowing-higher-returns

Bloomberg New Energy Finance (2013) 2030 Market Outlook. Bloomberg. http://www.enerpole.fr/uploads/news/id40/2030%20Market%20Outlook.pdf

Blue Smart Farms (2014) http://www.bluesmartfarms.com/home/

Bostrom N (2013) Existential Risk Prevention as Global Priority, Global Policy. http://www.existential-risk.org/concept.html

Box J (2014) Is the climate dragon awakening? June 2014. http://www.meltfactor.org/blog/?p=1329 and https://twitter.com/climate_ice

Brain CK (2009) A prehistoric detective story. Quest 5(2):15–19, Academy of Science of South Africa

Brain CK, Sillent A (1988) Evidence from the Swartkrans cave for the earliest use of fire. Nature 336:464–466. doi:10.1038/336464a0

Broecker WS (1975) Climatic change: are we on the brink of a pronounced global warming? Science 189(4201):460–463

Bro-Rasmussen F (1996) Contamination by persistent chemicals in food chain and human health. Sci Total Environ 188(Suppl):S45–S60. http://www.sciencedirect.com/science/article/pii/004896979605276X

Brown L (2004) Outgrowing the earth. Earth Policy Institute, Washington

Brown L (2011) Rising temperatures melting away Global Food Security. World on the Edge. http://www.earth-policy.org/book_bytes/2011/wotech4_ss3

Brown E (2015) Serving a thirsty world: trends in desalination. HIS Engineering, 8 January 2015. http://insights.globalspec.com/article/336/serving-a-thirsty-world-trends-in-desalination

Brummitt NA, Bachman SP, Griffiths-Lee J, Lutz M, Moat JF, Farjon A et al (2015) Green plants in the red: a baseline global assessment for the IUCN sampled red list index for plants. PLoS One 10(8), e0135152. doi:10.1371/journal.pone.0135152

Brunnstrom D (2015) US 'deeply concerned' by North Korean nuclear advances. Canberra Times. http://www.canberratimes.com.au/world/us-deeply-concerned-by-north-korean-nuclear-advances-20150225-13omi3.html. Accessed 25 Feb 2015

Buffet B, Archer D (2004) Global inventory of methane clathrate: sensitivity to changes in the deep ocean. Earth Planet Sci Lett 227:185–199

Bulletin of the Atomic Scientists (2015) It is 5 minutes to midnight. http://thebulletin.org/current-issue#. Accessed Jan 2015

Bulletin of the Atomic Scientists (2016) It is still 3 minutes to midnight. http://thebulletin.org/press-release/doomsday-clock-hands-remain-unchanged-despite-iran-deal-and-paris-talks9122

Calder J (undated) Persistent organic pollutants in the Arctic, NOAA. http://www.arctic.noaa.gov/essay_calder.html

California Department of Toxic Substances Control, Emerging Chemicals of Concern (2007) http://www.dtsc.ca.gov/assessingrisk/emergingcontaminants.cfm

Cameron D, Osborne C, Horton P, Sinclair M (2015) A sustainable model for intensive agriculture. University of Sheffield. http://grantham.sheffield.ac.uk/wp-content/uploads/2015/12/A4-sustainable-model-intensive-agriculture-spread.pdf. Accessed 2 Dec 2015

Campaign to Stop Killer Robots (2015) The problem. http://www.stopkillerrobots.org/the-problem/

Campbell K et al (2007) The Age of Consequences: the foreign policy and national security implications of Global Climate Change. Center for Strategic and International Studies

Campell K et al (2007) The age of consequences, CSIS. http://oai.dtic.mil/oai/oai?verb=getRecord&metadataPrefix=html&identifier=ADA473826

Carey B (2014) Stanford biologist warns of early stages of Earth's 6th mass extinction event. 24 July 2014. http://news.stanford.edu/pr/2014/pr-sixth-mass-extinction-072414.html

Carpenter DO, Arcaro K, Spink DC (2002a) Understanding the human health effects of chemical mixtures. Environ Health Perspect 100:259–269

Carpenter DO et al (2002b) Understanding the human health effects of chemical mixtures. Environ Health Perspect 110, February. http://www.ncbi.nlm.nih.gov/pmc/articles/PMC1241145/pdf/ehp110s-000025.pdf

Carter J (1999a) First step to peace is eradicating hunger. International Herald Tribune (17 June 1999)

Carter J (1999b) First step towards peace is eradicating hunger. New York Times. 17 June1999.http://www.nytimes.com/1999/06/17/opinion/17iht-edcarter.2.t.html

Carter E (2011) Dutch architecture firm rethinks the urban farm. ZDNet. http://www.smartplanet.com/blog/decoding-design/dutch-architecture-firm-rethinks-the-urban-farm/

Castelvecchi D, Schiermeier Q, Hodson R (2015) Can Islamic scholars change thinking on climate change? Nature, 19 August 2015. http://www.nature.com/news/can-islamic-scholars-change-thinking-on-climate-change-1.18203

Catton WR (1982) Overshoot. University of Illinois Press, Champaign

Ceballos G, Ehrlich PR, Barnosky AD, García A, Pringle RM, Palmer TM (2016) Accelerated modern human–induced species losses: entering the sixth mass extinction. Sci Adv 1:e1400253, 19 June 2015

Cellan-Jones R (2014) Stephen Hawking warns artificial intelligence could end mankind. BBC, December 2, 2014. http://www.bbc.com/news/technology-30290540

Centre for Biodiversity (2016) The extinction crisis. http://www.biologicaldiversity.org/programs/biodiversity/elements_of_biodiversity/extinction_crisis/

Chadha M (2014) World's largest coal miner to invest $1.2 billion in solar power. Renew Economy, 26 September 2014. http://reneweconomy.com.au/2014/worlds-largest-coal-miner-invest-1-2-billion-solar-power-39481

Chambers I, Humble J (2012) Plan for the planet: a business plan for a sustainable world. Gower, Farnham

Chatterjee P (2015) A chilling new post-traumatic stress disorder: why drone pilots are quitting in record numbers. Salon, March 7, 2015. http://www.salon.com/2015/03/06/a_chilling_new_post_traumatic_stress_disorder_why_drone_pilots_are_quitting_in_record_numbers_partner/

Children in History (2012) World War II: Japan and oil (1939–45). http://histclo.com/essay/war/ww2/stra/w2j-oil.html. Accessed 13 Nov 2013

Churchill WS (1947) UK House of Commons speech, 11 November 1947

Clark L (2012) Humans are made up of more microbes than human cells. Wired, 15 June 2012. http://www.wired.co.uk/news/archive/2012-06/15/human-microbes

Club of Rome (1972) Limits to growth. http://www.clubofrome.org/?p=326

Cohen L (1988) Everybody knows. Sony/ATV Music Publishing LLC, Universal Music Publishing Group

Colgan J (2013) Oil, conflict, and U.S. national interests. Policy Brief, Belfer Center for Science and International Affairs, Harvard Kennedy School. http://belfer-center.ksg.harvard.edu/publication/23517/oil_conflict_and_us_national_inter-ests.html. Accessed Oct 2013

Collins English Dictionary 10th Edition. 2014. HarperCollins Publishers. http://dictionary.reference.com/browse/wisdom

Connor S (2014) US scientist Professor Yoshihiro Kawaoka's mutated H1N1 flu virus 'poses a threat to human population if it should escape,' says critic. The Independent, 6 July 2014. http://www.independent.co.uk/news/science/us-scientist-professor-yoshihiro-kawaokas-mutated-h1n1-flu-virus-poses-a-threat-to-human-population-if-it-should-escape-says-critic-9587952.html

Conservation International (2015) The ocean. http://www.conservation.org/what/pages/oceans.aspx?gclid=Cj0KEQiA_ZOlBRD64c7-gOzvrP0BEi QAAYBndzvEe12GCnPRwrOv-ZIBqC2tSl7FaAbFQVvZxGLna3MaAhaG8P8 HAQ

Coughlan S (2013) How are humans going to become extinct? BBC, 24 April 2013, http://www.bbc.com/news/business-22002530

Council on Foreign Relations. Acc (2015) The Sunni-Shia divide. http://www.cfr.org/peace-conflict-and-human-rights/sunni-shia-divide/p33176#!/

Cox C et al (2012) Losing ground. EWG. http://www.ewg.org/losingground/report/executive-summary.html

Crawford J (2012) What if the world's soil runs out? Time. Dec 14, 2012. http://world.time.com/2012/12/14/what-if-the-worlds-soil-runs-out/. Accessed 12 Dec 2012

Credit Suisse Research Institute (2015) Global Wealth 2015. https://publications.credit-suisse.com/tasks/render/file/?fileID=F2425415-DCA7-80B8-EAD989AF9341D47E

Cribb JHJ (2001) The origin of acquired immune deficiency syndrome: can science afford to ignore it? Philos Trans R Soc Lond B Biol Sci 356(1410):935–938, http://www.ncbi.nlm.nih.gov/pmc/articles/PMC1088489/

Cribb JHJ (2007) The Dwarf Lords: tiny devices, tiny minds and the new enslave-ment. In: The Governance of Science and Technology, a Joint GovNet/

CAPPE/UNESCO conference, 9–10 August 2007, Australian National University, Canberra, Australia

Cribb JHJ (2011) The coming famine. University of California Press, Berkeley

Cribb JHJ (2012). Farm clearances at tipping point. Canberra Times, 23 July 2012

Cribb JHJ (2013) Food and fuel forever. Future Directions International. http://www.futuredirections.org.au/publication/food-and-fuel-forever/

Cribb JHJ (2014) Poisoned planet. Allen & Unwin, Australia. http://www.allenandunwin.com/default.aspx?page=94&book=9781760110468

Cribb JHJ (2016) Quantum computing and the dawn of the quantum tyranny. The Canberra Times, 10 January 2016. http://www.canberratimes.com.au/comment/dawn-of-the-quantum-tyranny-20160108-gm1tay

Crutzen PJ (2006) Earth system science in the anthropocene. Springer, Berlin

Cunningham S (2013) The creative economy could fuel Australia's next boom. The Conversation. http://theconversation.com/the-creative-economy-could-fuel-australias-next-boom-19108. Accessed 31 Oct 2013

Curnoe D (2013) Of heads and headlines: can a skull doom 14 human species? The Conversation. http://theconversation.com/of-heads-and-headlines-can-a-skull-doom-14-human-species-19227. Accessed 18 Oct 2013

Daily GC, Ehrlich P (1996) Impact of global development and change on the epidemiological environment. Environ Dev Econ 1(03):311–346, July 1996

Daily GC, Ehrlich P, Ehrlich A (1994) Optimum human population size. Popul Environ 15(6):469–475

Dal Toso R, Melandri F (2011) Sustainable sourcing of natural food ingredients by plant cell cultures. AgroFOOD Ind Hi-Tech 22(2)

Darimont CT et al (2015) The unique ecology of human predators. Science 349(6250):858–860. doi:10.1126/science.aac4249, http://www.sciencemag.org/content/349/6250/858.short

Dasgupta PS, Ehrlich PR (2013) Pervasive externalities at the population, consumption, and environment nexus. Science 340:324. doi:10.1126/science.1224664

de Chardin TP (1955) The phenomenon of man. Harper Perennial, New York

De Soya I, Gleditsch NP (1999) To cultivate peace: agriculture in a world of conflict. PRIO, Oslo

De Vos JM, Joppa LN, Gittleman JL, Stephens PR, Pimm SL (2014) Estimating the normal background rate of species extinction. Conserv Biol. doi:10.1111/cobi.12380

Dent B (2002) The hydrogeological context of cemetery operations and planning in Australia. University of Technology Sydney, Sydney

Despommier D (2013) The vertical farm. http://www.verticalfarm.com/

Diamond J (1993) The third chimpanzee. Harper Perennial, New York

Diamond J (2005) Collapse: how societies choose to fail or succeed. Viking, New York

Diamond JD (2006) Collapse: how societies choose to fail or succeed. Penguin, New York

Diaz RJ, Rosenberg R (2008) Spreading dead zones and consequences for marine ecosystems. Science 321(5891):926–929. doi:10.1126/science.1156401

Dietz R et al (2012) Three decades (1983–2010) of contaminant trends in East Greenland polar bears (*Ursus maritimus*). Environ Int 59(2012):485–493

Dirzo R et al (2014) Defaunation in the Anthropocene. Science 345:401

Discovery Newsletter (2013) Deserts spreading like 'cancer'. http://news.discovery.com/earth/deserts-middle-east.htm. Accessed 11 Feb 2013

Doll PJ et al (2012) Impact of water withdrawals from groundwater and surface water on continental water storage variations. Journal of Geodynamics 59–60:143–156

Drescher A (2005) Urban agriculture: a response to crisis. RUAF 2005. Also urban farming is growing a Green Future. National Geographic. http://environment.nationalgeographic.com.au/environment/photos/urban-farming/#/earth-day-urban-farming-new-york-rooftop_51631_600x450.jpg

Dunlop I (2014) Climate change—beyond dangerous. In: Goldie J, Betts K (eds) Sustainable futures: linking population, resources and the environment. CSIRO Publishing, Collingwood

Dyer G (2008) Climate wars. Scribe Publications, Melbourne

Dyer G (2009) Climate wars. Vintage, Canada

Dyke J (2016) Meltdown Earth: the shocking reality of climate change kicks in—but who is listening? The Conversation, 15 March 2016. https://theconversation.com/meltdown-earth-the-shocking-reality-of-climate change-kicks-in-but-who-is-listening-56255. For the actual data, see: http://data.giss.nasa.gov/gistemp/tabledata_v3/GLB.Ts+dSST.txt

ecoTECH (2012) ecoTECH will provide "beyond organic" vegetables and fish with power generated with a "net zero carbon emissions" Combined Heat and Power Station. 5 January 2012. http://www.ecotechenergygroup.com/index.php?mact=News,cntnt01,detail,0&cntnt01articleid=20&cntnt01origid=103&cntnt01returnid=109

Ehrlich P (1968) The population bomb. Ballantine Books, New York

Ehrlich PR, Ehrlich AH (2013) Can a collapse of global civilization be avoided? Proc R Soc B 280:20122845

Ehrlich PR, Ehrlich AH (2014) It's the numbers, stupid. In: Sustainable futures: linking population resources and the environment. CSIRO Publishing, Collingwood

Einstein A (1939/1941) Science and religion. http://www.westminster.edu/staff/nak/courses/Einstein%20Sci%20%26%20Rel.pdf

Eller D (2014) Erosion estimated to cost Iowa $1 billion in yield. Des Moines Register. 3 May 2014. http://www.desmoinesregister.com/story/money/agriculture/2014/05/03/erosion-estimated-cost-iowa-billion-yield/8682651/

Elliott L (2015) New Oxfam report says half of global wealth held by the 1%. The Guardian, 19 January 2015. http://www.theguardian.com/business/2015/jan/19/global-wealth-oxfam-inequality-davos-economic-summit-switzerland

Encyclopaedia Brittanica (2016) South sea bubble. https://www.britannica.com/event/South-Sea-Bubble

Encyclopaedia Britannica (2015) Taiping rebellion. https://www.britannica.com/event/Taiping-Rebellion

Environmental Working Group (2009) CDC analyzes toxics in humans. http://www.ewg.org/news/news-releases/2009/12/11/cdc-analyzes-toxics-humans

Environmental Working Group (2010) Cancer-causing chemical found in 89 percent of cities sampled. www.ewg.org/chromium6-in-tap-water

European Chemicals Agency (2015) http://echa.europa.eu/web/guest/information-on-chemicals/pre-registered-substances

Farnsworth T (2014) Background and status of Iran's nuclear program. Arms Control Association, Washington DC

Fey SB et al (2014) Recent shifts in the occurrence, cause, and magnitude of animal mass mortality events. Proc Natl Acad Sci 112(4). doi:10.1073/pnas.1414894112. http://www.pnas.org/content/112/4/1083.abstract

Fey SB et al (2015) Recent shifts in the occurrence, cause, and magnitude of animal mass mortality events. Proc Natl Acad Sci 112(4):1083–1088. doi:10.1073/pnas.1414894112

Finley B (2012) Colorado farms planning for dry spell losing auction bids for water to fracking projects. The Denver Post. 1 April 2012. http://www.dailycamera.com/boulder-county-news/ci_20299962/colorado-farms-planning-dry-spell-losing-auction-bids?source=rss

Fischer S (2015) Lithium extraction in the Chilean North, RedUse. http://www.reduse.org/en/blog/lithium-extraction-chilean-north

Fischetti M (2012) How much water do nations consume? Scientific American. http://www.scientificamerican.com/article/graphic-science-how-much-water-nations-consume/. Accessed 21 Mar 2012

Fisher M (2013) This alarming map shows dozens of radioactive materials thefts and losses every year. Washington Post. http://www.washingtonpost.com/blogs/worldviews/wp/2013/12/06/this-alarming-map-shows-dozens-of-nuclear-materials-thefts-and-losses-every-year/. Accessed 13 Dec 2013

Flannery T (2002) The future eaters. Grove Press, New York, First published 1994

Forbes (2014) The world's biggest public companies. Forbes, May. http://www.forbes.com/global2000/list/

Forest health (2015) Special issue of science. Science 349(6250):800–801. doi:10.1126/science.349.6250.800, http://www.sciencemag.org/content/349/6250/800.full. Accessed 21 Aug 2015

Foster P (2011) The top 10 Chinese food scandals. UK Daily Telegraph, 27 April 2011, http://www.telegraph.co.uk/news/worldnews/asia/china/8476080/Top-10-Chinese-Food-Scandals.html

French B (2016) Food Plants International. http://foodplantsinternational.com/

Fuoco R et al (2009) Persistent organic pollutants in the antarctic environment. Scientific Committee on Antarctic Research, Cambridge, www.scar.org/publications/occasionals/POPs_in_Antarctica.pdf

Fussell P (1989) Wartime: understanding and behavior in the second world war. Oxford University Press, New York

Galapagos Conservancy (2014). Lonesome George. http://www.galapagos.org/about_galapagos/lonesome-george/

Gencer EA (2013) Natural disasters, urban vulnerability, and risk management: a theoretical overview. Springer, Berlin

Gerland P, Raftery AE, Ševčíková H, Li N, Gu D, Spoorenberg T, Alkema L, Fosdick BK, Chunn J, Lalic N, Bay G, Buettner T, Heilig GK, Wilmoth J (2014) World population stabilization unlikely this century. Science 346(6206):234–237. doi:10.1126/science.1257469

Gilbert N (2009) The disappearing nutrient. Nature 461:716

Gillis J (2016) Zika outbreak could be an omen of the global warming threat. New York Times, 18 February 2016. http://www.nytimes.com/2016/02/19/science/zika-outbreak-could-be-an-omen-of-the-global-warming-threat.html?partner=rss&emc=rss&smid=tw-nytimesscience&smtyp=cur&_r=0

Glancy J (2014) EO Wilson: king of the ants has the gigantic task of saving us all. The Sunday Times, 9 November 2014. http://www.thesundaytimes.co.uk/sto/newsreview/features/article1480929.ece

Gleick P (2015) Water conflict chronology timeline. http://www2.worldwater.org/conflict/timeline/ and http://www2.worldwater.org/conflict/map/

Global Challenges Foundation (2015) 12 risks that threaten human civilisation, GCF 2015. http://globalchallenges.org/publications/globalrisks/about-the-project/

Global Counterterrorism Forum (2016) https://www.thegctf.org/web/guest/about

Global Energy Statistical Yearbook (2014) Enerdata. https://yearbook.enerdata.net/world-natural-gas-production.html

Global Footprint Network (2015) The ecological footprint of cities and regions: comparing resource availability with resource demand. Environ Urban 18(1): 103–112. http://www.footprintnetwork.org/en/index.php/GFN/page/footprint_for_cities/

Global Footprint Network (2016): At a glance. http://www.footprintnetwork.org/en/index.php/GFN/page/at_a_glance/

Global Nature Fund (2008) 13th world lakes conference. Wuhan, China. http://www.globalnature.org/30604/EVENTS/World-Lakes-Conference/02_vorlage.asp

Goodson WH et al (2015) Assessing the carcinogenic potential of low-dose exposures to chemical mixtures in the environment: the challenge ahead. Carcinogenesis 36(Suppl 1):S254-S296. doi:10.1093/carcin/bgv039, http://carcin.oxfordjournals.org/content/36/Suppl_1/S254.full

Grabianowski E (2015) How currency works. HowStuffWorks. http://money.howstuffworks.com/currency6.htm

Grandjean P, Landrigan PJ (2014) Neurobehavioural effects of developmental toxicity. Lancet. doi:10.1016/S1474-4422(13)70278-3

Grassini P, Eskridge KM, Cassman KG (2013) Distinguishing between yield advances and yield plateaus in historical crop production trends. Nat Commun. 17 December 2013. http://www.nature.com/ncomms/2013/131217/ncomms3918/full/ncomms3918.html

Greene J (2012) Detroit hospitals are growing, selling their own produce. Modern Healthcare. 17 September 2012. http://www.modernhealthcare.com/article/20120917/INFO/309179994

Grodzinski A (2011) Do video games influence violent behavior? Michigan Youth Violence Prevention Centre, 24 August 2011. http://yvpc.sph.umich.edu/2011/08/24/video-games-influence-violent-behavior/

Gubrud M (2015) Test ban for hypersonic missiles? Just say No. Bulletin of the Atomic Scientists. http://thebulletin.org/test-ban-hypersonic-missiles8422. Accessed 24 June 2015

Guiry D (2012) How many species of water plants are there? J Phycol 48:1057–1063

Gustavsson J, Cederberg C, Sonesson U (2011a) Global food waste and food losses: extent, causes and prevention. UNFAO. http://data.worldbank.org/indicator/EN.ATM.CO2E.PC

Gustavsson J, Cederberg C, Sonesson U (2011b) Global food losses and food waste: extent, causes and prevention. FAO, Rome. http://www.fao.org/docrep/014/mb060e/mb060e00.pdf

Haberl H et al (2007) Quantifying and mapping the human appropriation of net primary production in earth's terrestrial ecosystems. PNAS 104(31), July 31 2007. http://www.pnas.org/content/104/31/12942.full.pdf

Hagel C (2014) The Department of Defense must plan for the national security implications of climate change. White House, 13 October 2014. http://www.whitehouse.gov/blog/2014/10/13/defense-department-must-plan-national-security-implications-climate-change

Hamer M (1994) Deadly illusion brings death on roads. New Scientist, 18 June 1994. http://www.dougstewartonline.co.uk/pdfs/illusion.pdf

Hamilton C (2015) Geoengineering is no place for corporate profit making. The Guardian, February 18, 2015. http://www.theguardian.com/sustainable-business/2015/feb/17/geoengineering-is-no-place-for-corporate-profit-making

Hansen J (2009) Storms of my grandchildren. Bloomsbury, New York

Hansen J et al (2008) Target atmospheric CO2: where should humanity aim? Cornell University Library, Atmospheric and Oceanic Physics. http://www.columbia.edu/~jeh1/2008/TargetCO2_20080407.pdf

Hansen J et al (2013) Assessing "dangerous climate change": required reduction of carbon emissions to protect young people, future generations and nature. PLoS One, December 3

Hardoon D (2015) Wealth: having it all and wanting more. Oxfam, 19 January 2015. http://policy-practice.oxfam.org.uk/publications/wealth-having-it-all-and-wanting-more-338125

Harris S (2013) The roots of good and evil: an interview with Paul Bloom, 12 November 2013. http://www.samharris.org/blog/item/the-roots-of-good-and-evil

Hartman B (2011) Norway attack suspect had anti-Muslim, pro-Israel views. Jerusalem Post, 24 July 2011. http://www.jpost.com/International/Norway-attack-suspect-had-anti-Muslim-pro-Israel-views

Harvard Kennedy School (2015) The contested field of violent video games. 31 January 2015. http://journalistsresource.org/studies/government/criminal-justice/value-violent-video-games-research-roundup#

Heffer P, Prud'homme M (2014) Fertilizer outlook 2014-2018, International Fertilizer Industry Association (IFA), Paris

Hegelstad Ø (2014) pers com. http://www.miljogartneriet.no/

Held LE (2013) 8 New York City restaurants that grow their own food. Well+Good. http://www.wellandgoodnyc.com/2013/08/16/8-new-york-city-restaurants-that-grow-their-own-food/

Helfand (2015) Ira Helfand addresses 2014 Nobel peace laureates summit. https://www.youtube.com/watch?v=YBdfWhZekEA&feature=youtu.be. Accessed 5 Jan 2015

Hertsgard M (2000) Mikhail Gorbachev explains what's rotten in Russia. Salon. http://www.salon.com/2000/09/07/gorbachev/. Accessed 8 Sept 2000

Hettick L (2013) Cisco study projects 3.6 billion internet users by 2017. Network World Fusion, 31 May 2013

Hickey H (2014) World population to keep growing this century, hit 11 billion by 2100. UW Today. 18 September 2014. http://www.washington.edu/news/2014/09/18/world-population-to-keep-growing-this-century-hit-11-billion-by-2100/

Higham T et al (2014) The timing and spatio-temporal patterning of Neanderthal disappearance. Nature 512:306–309. doi:10.1038/nature13621, 21 August 2014

History of Money, Wikipedia, acc. (2015) http://en.wikipedia.org/wiki/History_of_money

Hoekstra AY, Chapagain AK (2007) Water footprints of nations: water use by people as a function of their consumption pattern. Water Resources Management 21(1):35–48

Hoekstra AY, Mekonnen MM (2011) The water footprint of humanity. PNAS 109(9):3232–3237. doi:10.1073/pnas.1109936109

Hopkins JNN (2012) The cloaca maxima and the monumental manipulation of water in archaic Rome. Institute of the Advanced Technology in the Humanities. Learning from Rome. 4 August 2012

Hou L, Zhang X, Wang D, Baccarelli A (2011) Effects of environmental chemicals on epigenetic changes. Int J Epidemiol 41(1):79–105. www.ncbi.nlm.nih.gov/pmc/articles/PMC3304523/#dyr154-B1

Howard BC (2015) Why did L.A. Drop 96 million 'Shade Balls' Into Its Water? National Geographic, 12 August 2015. http://news.nationalgeographic.com/2015/08/150812-shade-balls-los-angeles-California-drought-water-environment/

IBM (2015) IBM awarded IARPA Grant to advance research towards a universal quantum computer. IBM Newsroom, 8 December 2015. https://www-03.ibm.com/press/us/en/pressrelease/48258.wss

International Energy Agency (IEA) (2014) World energy outlook

IMDB (2012) Top 100 best disaster films. 14 March 2012. http://www.imdb.com/list/ls002913604/

Impacts. The lock-the-gate-alliance in Australia, http://www.lockthegate.org.au/impacts

Ingber S (2012) Lonesome George not the last of his kind after all. National Geographic, November 12, 2012

International Atomic Energy Agency (2014) AEA Incident and trafficking Database (ITDB), Incidents of nuclear and other radioactive material out of regulatory control. 2014 Fact Sheet. http://www-ns.iaea.org/downloads/security/itdb-fact-sheet.pdf

International Organisation of Motor Vehicle Manufacturers (OICA) (2013); Production Statistics. http://www.oica.net/category/production-statistics/

International Panel on Fissile Materials (2013) Global fissile materials report. http://fissilematerials.org/library/gfmr13.pdf

International Physicians for the Prevention of Nuclear War (2016) http://www.ippnw.org/

International Resources Panel (2011). Decoupling natural resource use and environmental impacts from economic growth. http://www.unep.org/resourcepanel/Portals/50244/publications/DecouplingENGSummary.pdf

International Resources Panel (2014) Decoupling 2, technologies, opportunities and policy options. http://www.unep.org/resourcepanel/AreasofResearchPublications/AssessmentAreasReports/Decoupling/tabid/133329/Default.aspx

International Rivers (2014) The state of the world's rivers. http://www.international-rivers.org/worldsrivers/. Accessed 25 Aug 2014

IPCC (2013) IPCC Fifth Assessment Synthesis Report—Summary for Decision Makers, November 2013. http://www.ipcc.ch/pdf/assessment-report/ar5/syr/SYR_AR5_SPMcorr2.pdf

IPCC (2014a) Summary for policymakers. https://www.ipcc.ch/pdf/assessment-report/ar5/syr/AR5_SYR_FINAL_SPM.pdf

IPCC (2014b) Fifth Synthesis Report. IPCC, Geneva.

IUCN (2016) http://iucn.org/

Jackson MO, Morelli M (2011) The reasons for wars—an updated survey. In: Coyne C (ed) Handbook on the political economy of war. Elgar Publishing, Cheltenham

Jasny L, Waggle J, Fisher DR (2015) An empirical examination of echo chambers in US climate policy networks. Nat Clim Chang. doi:10.1038/NCLIMATE2666

Jones S (2014) Church of England vows to fight 'great demon' of climate change. The Guardian, 13 February 2014, http://www.theguardian.com/world/2014/feb/12/church-climate-change-investment-great-demon-flooding

Kanter J (2009) Scientist: warming could cut population to 1 billion. New York Times. http://dotearth.blogs.nytimes.com/2009/03/13/scientist-warming-could-cut-population-to-1-billion/

Kaplan S (2015) The disgusting 10-ton 'fatberg' that broke a London sewer. Washington Post, April 22, 2015. http://www.washingtonpost.com/news/morning-mix/wp/2015/04/22/the-disgusting-ten-ton-fatberg-that-broke-a-london-sewer/

Kellenberger J (2010) Bringing the era of nuclear weapons to an end. ICRC. https://www.icrc.org/eng/resources/documents/statement/nuclear-weapons-statement-200410.htm. Accessed 20 Apr 2010

Kennett JP, Cannariato KG, Hendy IL, Behl RJ (2003) Methane hydrates in quaternary climate change: the clathrate gun hypothesis. American Geophysical Union, Washington. ISBN 0-87590-296-0

Ker P (2014) 'Green horizon' may force BHP to quit coal, says Andrew Mackenzie. Sydney Morning Herald, November 25, 2014. http://www.smh.com.au/business/mining-and-resources/green-horizon-may-force-bhp-to-quit-coal-says-andrew-mackenzie-20141125-11t74d.html

Kerr PK (2008) Nuclear, biological, and chemical weapons and missiles: status and trends. CRS Report for Congress. Congressional Research Service (CRS). https://www.fas.org/sgp/crs/nuke/RL30699.pdf. Accessed 20 Feb 2008

Khanna P (2013) The end of the nation-state? New York Times, 12 October 2013

Kirmayer LJ et al (2004) Explaining medically unexplained symptoms. Can J Psychiatry 49(10):663–672

Klein DR (1966) The introduction, increase and crash of reindeer on St Matthew Island. University of Alaska. http://dieoff.org/page80.htm

Klein N (2014) This changes everything: capitalism vs. the climate. Simon & Schuster, New York

Klotz LC, Sylvester E (2012) The unacceptable risks of a man-made pandemic. Bull At Sci. http://thebulletin.org/unacceptable-risks-man-made-pandemic

Kobylewski S, Jacobson MF (2010) Food dyes: a rainbow of risks. Center for Science in the Public Interest. https://www.cspinet.org/fooddyes/

Kolasanti KJA et al (2012) The city as an "agricultural powerhouse"? Perspectives on expanding urban agriculture from Detroit, Michigan. Urban Geogr 33(3). http://www.tandfonline.com/doi/abs/10.2747/0272-3638.33.3.348#.VP-l5VIcSdM

Kolbert E (2014) The sixth extinction. Bloomsbury, London

Konkel L (2012) Antarctic wilds carry as much chemical flame retardants as urban rivers. Scientific American, February 12, 2014. http://www.scientificamerican.com/article/antarctic-wilds-carry-as-much-chemical-flame-retardants-as-urban-rivers/

Koronowski R (2016) Record-breaking Hot Ocean temperatures are frying the Great Barrier Reef. ThinkProgress. http://thinkprogress.org/climate/2016/04/26/3769440/great-barrier-reef-bleaching/, 26 April 2016

Krauss L (2013) Deafness at doomsday. New York Times. http://www.nytimes.com/2013/01/16/opinion/deafness-at-doomsday.html?_r=2&. Accessed 15 Jan 2013

Kravcik M et al (2008) Water for the recovery of the climate: a new water paradigm. People and Water NGO. www.waterparadigm.org

Kriger KM, Hero J-M (2009) Chytridiomycosis, amphibian extinctions, and lessons for the prevention of future panzootics. EcoHealth. doi:10.1007/s10393-009-0228-y

Kristensen HM, Norris RS (2013) Israeli nuclear weapons. Global nuclear weapons inventories, 1945–2013. Bull At Sci 69(5):75–81, September/October 2013

Kuhlman A (2015) Peak Oil. http://www.oildecline.com/

Kuroda K, Fukushi T (2008) Groundwater contamination in urban areas, groundwater management in Asian cities. Springer, Tokyo

Lakenet (2015) World Lakes Database. http://www.worldlakes.org/index.asp

Lakepedia (2015) Lake Chad: the shrinking giant. http://www.lakepedia.com/lake/chad.html

Leakey R (1996) The sixth extinction. Random House, New York

Legg-Bagg G (2014) Blue Smart Farms. http://www.bluesmartfarms.com

Lehman E (2015) Extreme rain may flood 54 million people by 2030. Scientific American. 5 March 2015. http://www.scientificamerican.com/article/extreme-rain-may-flood-54-million-people-by-2030/?utm_source=twitterfeed&utm_medium=twitter&utm_campaign=Feed%3A+ScientificAmerican-Twitter+%28Content%3A+Global+Twitter+Feed%29

Lenntech (2014) Use of water in food and agriculture. http://www.lenntech.com/water-food-agriculture.htm

Levi M (2007) On nuclear terrorism. Harvard University Press, Cambridge, MA

Lewis HH (1959) The Great Case of The Canal vs. The Railroad—4 Gill & Johnson 1 (1832), 19 Md. L. Rev. 1 http://digitalcommons.law.umaryland.edu/mlr/vol19/iss1/3

Liesowska A, Lambie D (2014) How global warming could turn Siberia into a giant crater 'time bomb'. The Siberian Times, December 25, 2014. http://siberiantimes.com/science/casestudy/news/n0076-how-global-warming-could-turn-siberia-into-a-giant-crater-time-bomb/

Linebaugh H (2013) I worked on the US drone program. The public should know what really goes on. The Guardian, December 29 2013. http://www.theguardian.com/commentisfree/2013/dec/29/drones-us-military

Loganathan BG, Kwan-Sing Lam P (2014) Global contamination trends of persistent organic chemicals. CRC Press 2011 or Global Contamination Initiative, CRC CARE, 2014, http://www.crccare.com/files/dmfile/CRCCAREGRCI brochure2.pdf

Lomborg B (2001) Including, for instance. In: The sceptical environmentalist. Cambridge University Press, Cambridge

Lordkipanidze D et al (2013) Complete skull from Dmanisi, Georgia, and the evolutionary biology of early homo. Science 342(6156):326–331

Lovelock J (2009) The vanishing face of Gaia: a final warning. Allen Lane, London

Maastricht University (2013) First-ever public tasting of lab-grown Cultured Beef burger. 5 August 2013

Mahasneh HI (2003) Humans and the environment: Islamic faith statement, faith in conservation. World Bank. http://www.arcworld.org/faiths.asp?pageID=75

Mair P (2014) Ruling the void: the hollowing of western democracy. Verso, July 2014

Malm O (1998) Gold mining as a source of mercury exposure in the Brazilian Amazon. Environ Res 77(2):73–78

Manikkam M et al (2013) Plastics derived endocrine disruptors (BPA, DEHP and DBP) induce epigenetic transgenerational inheritance of obesity, reproductive disease and sperm epimutations. PLoS One 8(1):e55387

Margolis J (2012) Growing food in the desert: is this the solution to the world's food crisis? The Guardian. 25 November 2012. http://www.theguardian.com/environment/2012/nov/24/growing-food-in-the-desert-crisis

Margulis L, Sagan D (1986) Microcosmos: four billion years of evolution from our microbial ancestors. University of California Press, Berkeley

Markey PM et al (2014) Violent video games and real-world violence: rhetoric versus data. Psychology of Popular Media Culture, 18 August 2014. http://psycnet.apa.org/psycinfo/2014-33466-001/

Marler JB, Wallin JR (2006) Human health, the nutritional quality of harvested food and sustainable farming systems. Nutrition Security Institute, USA

Matthews JA, Tan H (2014) China's renewable energy revolution: what is driving it? Asia-Pac J 12(44), No. 3, November 3, 2014. http://www.japanfocus.org/-Hao-Tan/4209

McCaulay DJ et al (2015) Marine defaunation: animal loss in the global ocean. Science 347(6219). doi:10.1126/science.1255641. http://www.sciencemag.org/content/347/6219/1255641

McCloskey B et al (2014) Emerging infectious diseases and pandemic potential: status quo and reducing risk of global spread. Lancet. http://www.thelancet.com/journals/laninf/article/PIIS1473-3099(14)70846-1/fulltext

McGlade C, Elkins P (2015) The geographical distribution of fossil fuels unused when limiting global warming to 2°C. Nature, 8 January 2015. http://www.nature.com/nature/journal/v517/n7533/full/nature14016.html

McKenzie FC, Williams J (2015) Sustainable food production: constraints, challenges and choices by 2050. Food Security. doi:10.1007/s12571-015-0441-1

McMichael AJ (2012) Insights from past millennia into climatic impacts on human health and survival. PNAS, March 27, 2012. http://www.ncbi.nlm.nih.gov/pmc/articles/PMC3324023/

Mekonnen MM, Hoekstra AY (2016) Four billion people facing severe water scarcity. Sci Adv 2(2), e1500323, http://advances.sciencemag.org/content/2/2/e1500323

Messer EM et al (1998) Food from peace: breaking the links between conflict and hunger. International Food Policy Research Institute, brief no 50, June 1998

Messerschmidt M (1990) Foreign policy and preparation for war. In: Germany and the second world war, vol. 1. Clarendon Press, Oxford

Meyerson B (2015) Top 10 emerging technologies of 2015. Scientific America, 4 March 2015. http://www.scientificamerican.com/article/top-10-emerging-technologies-of-20151/

Mian AR, Sufi A, Trebbi F (2012) Resolving debt overhang: political constraints in the aftermath of financial crises. NBER Working Paper No. 17831, February 2012

Minerals Council of Australia (2013) MCA Gender Diversity White Paper. MCA. http://www.minerals.org.au/file_upload/files/resources/education_training/MCA_Gender_Diversity_White_Paper_Summary_FINAL.PDF

Monastersky R (2014) Life—a status report. Nature 516:158–161. doi:10.1038/516158a, 10 December 2014

Montgomery DR (2007) Dirt: the erosion of civilizations. University of California Press, Berkeley, CA

Moon BK (2015) We are the last generation that can fight climate change. We have a duty to act. The Guardian, Monday 12 January 2015. http://www.theguardian.com/commentisfree/2015/jan/12/last-generation-tackle-climate-change-un-international-community

Moore C (2007) Six escalation scenarios leading to nuclear war. http://www.carol-moore.net/nuclearwar/alternatescenarios.html

Mora C, Tittensor DP, Adl S, Simpson AGB, Worm B (2011) How many species are there on earth and in the ocean? PLoS Biol 9(8), e1001127. doi:10.1371/journal.pbio.1001127

Mössner S, Ballschmiter K (1997) Marine mammals as global pollution indicators for organochlorines. Chemosphere 34(5–7):1285–1296

Motherboard (2014) For good or bad, intelligent, swarming nanobots are the next frontier of drones. Motherboard, 21 May 2014. http://motherboard.vice.com/read/why-the-us-military-is-funding-tiny-autonomous-flying-robots

Motherboard (2016) Bee extinction is threatening the world's food supply, UN warns. Motherboard, 27 February 2016. http://motherboard.vice.com/read/bee-extinction-is-threatening-the-worlds-food-supply-un-warns?utm_source=mbtwitter

Muir DCG et al (2002) Toxaphene and other persistent organochlorine pesticides in three species of albatrosses from the North and South Pacific Ocean. Environ Toxicol Chem 21(2):413–423

Muncke J, Peterson Myers J, Scheringer M, Porta M (2014) Food packaging and migration of food contact materials: will epidemiologists rise to the neotoxic challenge? J Epidemiol Community Health. doi:10.1136/jech-2013-202593

Musser D, Sunderland D (2005) War or words: interreligious dialogue as an instrument of peace Cleveland. The Pilgrim Press, Cleveland

NASA (2015) Quantum Artificial Intelligence Laboratory. http://www.nas.nasa.gov/quantum/

National Water Commission (2008) Emerging trends in desalination: a review. http://www.nwc.gov.au/__data/assets/pdf_file/0009/11007/Waterlines_-_Trends_in_Desalination_-_REPLACE_2.pdf

Natural Resources Institute of Finland (2015) Global warming reduces wheat production markedly if no adaptation takes place. 12 January 2015. http://www.luke.

fi/en/tiedote/global-warming-reduces-wheat-production-markedly-if-no-adaptation-takes-place/

Naylor D (2014) Carl Linnaeus ranked most influential person of all time. Uppsala University. http://www.uu.se/en/media/news/article/?id=3519&area=2,7,16&typ=artikel&na=&lang=en

Newbold T et al (2016) Has land use pushed terrestrial biodiversity beyond the planetary boundary? A global assessment. Science 353(6296):288–291

Necrometrics (2012) Selected death tolls for wars, massacres and atrocities before the 20th century. http://necrometrics.com/pre1700a.htm

New Scientist (2016) Many of world's lakes are vanishing and some may be gone forever. https://www.newscientist.com/article/2079562-many-of-worlds-lakes-are-vanishing-and-some-may-be-gone-forever/?utm_source=&utm_medium=&utm_campaign=. Accessed 4 Mar 2016

News Ltd. (2014) Is the US at risk of a hyperinflation collapse? News.com.au, 4 December 2014. http://www.news.com.au/finance/money/is-the-us-at-risk-of-a-hyperinflation-collapse/story-e6frfmci-1227144167041

NOAA (2014) Global analysis—annual 2014. http://www.ncdc.noaa.gov/sotc/global/2014/13

Northeast Fisheries Science Centre (2008) Persistent man-made chemical pollutants found in deep-sea octopods and squids, June 2008, www.nefsc.noaa.gov/press_release/2008/SciSpot/ss0810

Nuclear Information and Resource Service (1996) Nuclear power plant fuel—a source of plutonium for weapons? http://www.nirs.org/factsheets/plutbomb.htm

OECD (1996) The knowledge-based economy. OECD, Paris

OECD (2015) Material resources, productivity and the environment. http://www.oecd.org/env/waste/material-resources-productivity-and-environment.htm

Onstot J, Ayling R, Stanley J (2010) Characterization of HRGC/MS unidentified peaks from the analysis of human adipose tissue. United States Environmental Protection Agency, June 30, 1987.

Oxford Dictionaries (2016) http://www.oxforddictionaries.com/definition/english/belief

Paleczny M, Hammill E, Karpouzi V, Pauly D (2015) Population trend of the world's monitored seabirds, 1950–2010. PLoS One 10(6), e0129342. doi:10.1371/journal.pone.0129342

Pamlin D et al (2015) 12 risks that threaten human civilization. Global Challenges Foundation, February 2015. http://globalchallenges.org/wp-content/uploads/12-Risks-with-infinite-impact-Executive-Summary.pdf

Papadoupolou S (2014) First observations of methane release from Arctic Ocean hydrates, SWERUS-C3. http://www.swerus-c3.geo.su.se/index.php/swerus-c3-in-the-media/news/177-swerus-c3-first-observations-of-methane-release-from-arctic-ocean-hydrates

Parker G (2013) Global crisis: war, climate change and catastrophe in the seventeenth century. Yale University Press, New Haven. http://www.amazon.com/Global-Crisis-Climate-Catastrophe-Seventeenth/dp/0300153236

Patterson R (2014) A closer look at Saudi Arabia. Peak Oil Barrel. 27 May 2014. http://peakoilbarrel.com/closer-look-saudi-arabia/

Pauly D, Zeller D (2016) Catch reconstructions reveal that global marine fisheries catches are higher than reported and declining. Nat Commun. 19 January 2016. http://www.nature.com/ncomms/2016/160119/ncomms10244/full/ncomms10244.html

Pearce F (2006) When the rivers run dry: water--the defining crisis of the twenty-first century. Beacon, Boston

Pearce F (2011) Phosphate: a critical resource - misused and now running low, Yale. 7 July 2011. http://e360.yale.edu/feature/phosphate_a_critical_resource_misused_and_now_running_out/2423/

Perkins S (2013) Oldest primate skeleton unveiled. Nature, 5 June 2013. http://www.nature.com/news/oldest-primate-skeleton-unveiled-1.13142

Peryman L (2012) Unchecked industry reduces land of a thousand lakes to a struggling few. Probe Int. http://journal.probeinternational.org/2012/07/20/unchecked-industry-reduces-land-of-a-thousand-lakes-to-a-struggling-few/. Accessed 20 July 2012

Pew Research Centre (2015) The future of world religions: population growth projections, 2010-2050, 2 April 2015. http://www.pewforum.org/2015/04/02/religious-projections-2010-2050/

Piketty T (2014) Capital in the twenty-first century. Harvard University Press, Cambridge

Plantagon (2016) Urban industrial vertical farming. http://plantagon.com/urban-agriculture/vertical-greenhouse

Plato. C 360BCE. The Republic: allegory of the cave

Pope A (2014) Open secrets: Julian Assange, Chelsea Manning, Edward Snowden and the role of the individual in challenging the War on Terror, University of Glasgow, February 2014. http://www.academia.edu/9716019/Open_Secrets_Julian_Assange_Chelsea_Manning_Edward_Snowden_and_the_role_of_the_individual_in_challenging_the_War_on_Terror

Pope Francis (2015) Encyclical Letter of the Holy Father Franciscus on Care for Our Common Home, May 24, 2015. http://w2.vatican.va/content/francesco/en/encyclicals/documents/papa-francesco_20150524_enciclica-laudato-si.html

Porta M, Lee DH (2012) Review of the science linking chemical exposures to the human risk of obesity and diabetes. ChemTrust UK, March 2012, www.wecf.eu/download/2012/March/CHEMTrustObesityDiabetesSummaryReport.pdf

Porter E (2014) The politics of income inequality. New York Times, 13 May 2014

PriceWaterhouseCoopers (2012) World in 2050: the BRICs and beyond: prospects, challenges and opportunities

Provieri F, Pirrone N (2005) Mercury pollution in the Arctic and Antarctic regions: dynamics of mercury pollution on regional and global scales. Springer, New York

Projection Based on OECD (2015) Material resources, productivity and the environment. http://www.oecd.org/env/waste/material-resources-productivity-and-environment.htm

Prufer K et al (2014) The complete genome sequence of a Neanderthal from the Altai mountains. Nature 505:33–39. doi:10.1038/nature12886, 02 January 2014

PSI (2015) PS21 survey: experts see increased risk of nuclear war. Project for study of the 21st century. https://projects21.org/2015/11/12/ps21-survey-experts-see-increased-risk-of-nuclear-war/. Accessed 12 Nov 2015

Quobil R (2015) Waiting for the Sea. BBC. http://www.bbc.com/news/resources/idt-a0c4856e-1019-4937-96fd-8714d70a48f7

Rahmstorf S (2013) Paleoclimate: the end of the Holocene. RealClimate. 16 September 2013. http://www.realclimate.org/index.php/archives/2013/09/paleoclimate-the-end-of-the-holocene/

Ramphal S (1992) Our country, the planet. Lime Tree Press, London

Ramsey L (2015) These 10 cities have the worst air pollution in the world, and it is up to 15 times dirtier than what is considered healthy. Business Insider, 21 September 2015. http://www.businessinsider.com.au/these-are-the-cities-with-the-worst-air-pollution-in-the-world-2015-9

Ratcliffe R (2015) 10-tonne fatberg removed from west London sewer. The Guardian, 22 April 2015. http://www.theguardian.com/uk-news/2015/apr/21/huge-10-ton-fatberg-removed-chelsea-sewer-london

Raup DM (1986) Biological extinction in earth history. Science 231(4745):1528–1533. doi:10.1126/science.11542058

Reagan R (1984) Best Reagan quotes on nuclear weapons. http://www.thereaganvision.org/quotes/

Regalado A (2016) Top U.S. intelligence official calls gene editing a WMD threat, MIT Technology Review, February 9, 2016

Rees M (2004) Our final century. Arrow Books. http://www.amazon.com/Our-Final-Century-Humanitys-Survival/dp/0099436868/ref=pd_sim_14_1?ie=UTF8&dpID=41FXHRDf6bL&dpSrc=sims&preST=_AC_UL160_SR105%2C160_&refRID=0PA1Y918F64N5QQVEGVW

Renewables (2014) Global status report. http://www.ren21.net/REN21Activities/GlobalStatusReport.aspx. Accessed 24 June 2014

Resource Efficiency: Economics and Outlook for Asia and the Pacific (REEO) (2011) United national environment program. http://www.unep.org/dewa/Portals/67/pdf/Resource_Efficiency_EOAP_web.pdf

Rice D (2015) Doomsday ticks closer to midnight: 'the probability of global catastrophe is very high'. Canberra Tines. http://www.canberratimes.com.au/environment/climate-change/doomsday-ticks-closer-to-midnight-the-probability-of-global-catastrophe-is-very-high-20150122-12wd4s.html. Accessed 23 Jan 2015

Rios L, Moore C, Jones PR (2007) Persistent organic pollutants carried by synthetic polymers in the ocean environment. Mar Pollut Bull 54(8):1230–1237. www.sciencedirect.com/science/article/pii/S0025326X07001324

Robertson E, Pinstrup-Andersen P (2010) Global land acquisition: neo-colonialism or development opportunity? Food Security 2(3):271–283

Robock A (2009) Nuclear winter. Encyclopaedia of earth. http://www.eoearth.org/view/article/154973/

Robock A, Toon CB (2012) Local nuclear war, global suffering. Chapter 4.3 in Lights out: how it all ends. Sci Am 2012

Rockström J, Steffen W, Noone K, Persson Å, Chapin FS III, Lambin E, Lenton TM, Scheffer M, Folke C, Schellnhuber H, Nykvist B, De Wit CA, Hughes T, van der Leeuw S, Rodhe H, Sörlin S, Snyder PK, Costanza R, Svedin U, Falkenmark M, Karlberg L, Corell RW, Fabry VJ, Hansen J, Walker B, Liverman D, Richardson K, Crutzen P, Foley J (2009) Planetary boundaries: exploring the safe operating space for humanity. Ecol Soc 14(2):32, http://www.ecologyandsociety.org/vol14/iss2/art32/ [online]

Rohde RA, Muller RA (2015) Air pollution in China: mapping of concentrations and sources. Berkeley Earth, May 2015, http://berkeleyearth.org/wp-content/uploads/2015/08/China-Air-Quality-Paper-July-2015.pdf

Romero S (2015) Taps start to run dry in Brazil's largest city. New York Times, 16 February, 2015. http://www.nytimes.com/2015/02/17/world/americas/drought-pushes-sao-paulo-brazil-toward-water-crisis.html?_r=1

Rosenfeld E (2014) Elon Musk's deleted message: five years until 'dangerous' AI. CNBC, 17 November 2014. http://www.cnbc.com/id/102192439#

Rosny J-H (1911) La Guerre du Feu. Editions Fasquelle.

Ross PS et al (2004) Harbor seals (*Phoca vitulina*) in British Columbia, Canada, and Washington State, USA, reveal a combination of local and global polychlorinated biphenyl, dioxin, and furan signals. Environ Toxicol Chem 23(1):157–165

Ruiz R (2010) Industrial chemicals lurking in your bloodstream. Forbes Magazine, 21 January 2010, www.forbes.com/2010/01/21/toxic-chemicals-bpa-lifestyle-health-endocrine-disruptors.html

Russell S et al (2015) Research priorities for robust and beneficial artificial intelligence: an open letter. Future of Life Institute, January 2015. http://futureoflife.org/misc/open_letter

Sachs JD (2015a) The war with radical Islam. Project syndicate. http://www.project-syndicate.org/commentary/radical-islam-western-military-intervention-by-jeffrey-d-sachs-2015-01. Accessed 15 Jan 2015

Sachs J (2015b) By separating nature from economics, we have walked blindly into tragedy. The Guardian, 10 March 2015, http://www.the-guardian.com/global-development-professionals-network/2015/mar/10/jeffrey-sachs-economic-policy-climate-change?CMP=share_btn_link

Santini J-L (2015) Climate change brings world closer to 'doomsday', say scientists. Agence France Press. http://www.smh.com.au/environment/climate-change/climate-change-brings-world-closer-to-doomsday-scientists-say-20150122-12wex9.html. Accessed 23 Jan 2015

Schiffman R (2013) Hunger, food security, and the African land grab. Ethics & International Affairs 27:239–249. doi:10.1017/S0892679413000208

Schiller B (2014) Floating ocean greenhouses bring fresh food closer to megacities. Co-Exist. 7 July 2014. http://www.fastcoexist.com/3032302/floating-ocean-greenhouses-bring-fresh-food-closer-to-megacities

Schlenker W, Roberts MJ (2009) Nonlinear temperature effects indicate severe damages to U.S. crop yields under climate change. PNAS 106(37). http://www.pnas.org/content/106/37/15594.short

Schlenker W, Roberts MJ (2009) Nonlinear temperature effects indicate severe damages to U.S. crop yields under climate change. PNAS 106(37). http://www.pnas.org/content/106/37/15594.short   http://www.nature.com/nature/journal/v427/n6970/abs/nature02121.html

Schnellnhuber HJ (2009) http://dotearth.blogs.nytimes.com/2009/03/13/scientist-warming-could-cut-population-to-1-billion/?_r=1

Scholes MC, Scholes RJ (2013) Dust unto dust. Science 342(6158):565. doi:10.1126/science.1244579

Schramski J et al (2015) Human domination of the biosphere: rapid discharge of the earth-space battery foretells the future of humankind. Proceedings of the National Academy of Sciences. www.pnas.org/cgi/doi/10.1073/pnas.1508353112 and http://phys.org/news/2015-07-destruction-earth-life-humans-jeopardy.html#jCp

Schuster-Wallace CJ, Sandford R (2015a) Water in the world we want. United Nations University Institute for Water. Environment and Health. http://inweh.unu.edu/wp-content/uploads/2015/02/Water-in-the-World-We-Want.pdf

Schuster-Wallace CJ, Sandford R (2015b) Water in the world we want. United Nations University Institute for Water, Environment and Health. http://inweh.unu.edu/wp-content/uploads/2015/02/Water-in-the-World-We-Want.pdf   and United Nations Office for Sustainable Development

Schuur EAG et al (2015) Climate change and the permafrost carbon feedback. Nature 520:171–179. doi:10.1038/nature14338. http://www.nature.com/nature/journal/v520/n7546/full/nature14338.html

Schwartz JH (2000) Taxonomy of the Dmanisi Crania. Science 289(5476):55–56

ScienceDaily (2014) BMJ—British Medical Journal: "Food packaging chemicals may be harmful to human health over long term." ScienceDaily, 19 February 2014. www.sciencedaily.com/releases/2014/02/140219205215.htm

Scott JM (2008) Threats to biological diversity: global, continental, local. U.S. Geological Survey, Idaho Cooperative Fish and Wildlife, Research Unit, University of Idaho

Seafish (UK) (2013) Contaminants. www.seafish.org/industry-support/legislation/contaminants

Segers H (2009) Introduction to scientific nomenclature. Hue University, Vietnam

Semeena VS, Lammel G (2005) The significance of the grasshopper effect on the atmospheric distribution of persistent organic substances. Geophys Res Lett 32(7). doi:10.1029/2004GL022229

Sharp R (2009) CDC scientists find rocket fuel chemical in infant formula, EWG, 2 April 2009. http://www.ewg.org/research/cdc-scientists-find-rocket-fuel-chemical-infant-formula

Shortell D (2015) Marcy Borders, survivor known as 'Dust Lady' in iconic 9/11 photo, dies at 42, CNN, August 27, 2015. http://edition.cnn.com/2015/08/26/us/9-11-survivor-dust-lady-dies/

Smithsonian Institution (2015) Ocean portal: ocean acidification. http://ocean.si.edu/ocean-acidification?gclid=Cj0KEQiA_ZOlBRD64c7-gOzvrP0BEiQAAYBndz4CUncsCFZfke02BK5q_id5kPPq7b_aJ1U49_1G-7kaAmSh8P8HAQ

Snow D, Hannam P (2014) Climate change could make humans extinct, warns health expert. Sydney Morning Herald, 31 March 2014. http://www.smh.com.au/environment/climate-change/climate-change-could-make-humans-extinct-warns-health-expert-20140330-35rus.html

Snowden E (2015) Edward Snowden, ABC Big Ideas, 19 May 2015. http://www.abc.net.au/radionational/programs/bigideas/edward-snowden-_-on-mass-surveilance/6464576

Steffen W, Crutzen PJ, McNeill JR (2007) The anthropocene: are humans now overwhelming the great forces of nature. Ambio 36(8):614–621. doi:10.1579/0044-7447(2007)36[614:TAAHNO]2.0.CO;2

Steffen W, Rockstrom J et al (2009) Planetary boundaries: guiding human development on a changing planet. Science 347(6223). doi: 10.1126/science.1259855. http://science.sciencemag.org/. (13 Feb 2015)

Steffen W, Rockstrom J et al (2015) Planetary boundaries: guiding human development on a changing planet. Science 347(6223). doi:10.1126/science.1259855

Stehle S, Schulz R (2015) Agricultural insecticides threaten surface waters at the global scale. Proc Acad Sci. http://www.pnas.org/content/early/2015/04/08/1500232112.long

Stern N (2006) Review on the economics of climate change, UK Government, 2006. http://siteresources.worldbank.org/INTINDONESIA/Resources/226271-1170911056314/3428109-1174614780539/SternReviewEng.pdf

Stiglitz J (2016) The new generation gap. Project Syndicate, 16 March 2016. https://www.project-syndicate.org/commentary/new-generation-gap-social-injustice-by-joseph-e--stiglitz-2016-03

Stockholm Convention (2013a) http://chm.pops.int/Convention/ThePOPs/The12InitialPOPs/tabid/296/Default.aspx and http://chm.pops.int/Implementation/NewPOPs/TheNewPOPs/tabid/672/Default.aspx

Stockholm Convention (2013b) Results of the global survey on concentrations in human milk of persistent organic pollutants by the United Nations Environment Programme and the World Health Organization, Stockholm Convention

Report, Geneva, May 2013, http://www.google.com.au/url?sa=t&rct=j&q=&esr c=s&source=web&cd=6&ved=0CEEQFjAF&url=http%3A%2F%2Fchm.pops. int%2FPortals%2F0%2Fdownload.aspx%3Fd%3DUNEP-POPS-COP.6-INF-33.English.pdf&ei=3GvlVN2OMcyl8AWt-IDoDA&usg=AFQjCNFs33wUC5 Qnc6zA9TnzPe9S1kj8Wg&sig2=3QLA7vFpouPcCjLSLgt6aA

Stockholm International Peace Research Institute (2014) SIPRI yearbook 2014

Stockholm International Peace Research Institute (SIPRI) (2016) World military spending resumes upward course, says SIPRI. Press release. http://www.sipri.org/ media/pressreleases/2016/milex-apr-2016. Accessed 5 Apr 2016

Stoycheff E (2016) Under surveillance: examining Facebook's spiral of silence effects in the wake of NSA internet monitoring. J Mass Commun Q, 1, http://m.jmq. sagepub.com/content/early/2016/02/25/1077699016630255.full.pdf?ijkey=1jxr Yu4cQPtA6&keytype=ref&siteid=spjmq

Surviving Earth (2014) Bindi Irwin quoted in the documentary film by Peter Charles Downey. http://www.survivingearthmovie.com/

Swamy V (2015) What percentage of the world's money is digital? Quora. http:// www.quora.com/What-percentage-of-the-worlds-money-is-digital

Talbot D (2014) Desalination out of desperation. MIT Technol Rev. http://www. technologyreview.com/featuredstory/533446/desalination-out-of-desperation/

The Doomsday Clock (1984) Southeast Missourian. Accessed 22 Feb 1984

The Economist (2013) Crash course: the origins of the Global Financial Crisis. The Economist. 7 September 2013. http://www.economist.com/news/ schoolsbrief/21584534-effects-financial-crisis-are-still-being-felt-five-years-article

The Economist (2014) Reservoir hogs: government responded late to a drought in Brazil's industrial heartland. The Economist, December 20, 2014. http://www. economist.com/news/americas/21636782-government-responded-late-drought-brazils-industrial-heartland-reservoir-hogs

The Economist (2015a) Barbarians at the farm gate. The Economist. 3 January 2015. http://www.economist.com/news/finance-and-economics/21637379-hardy investors-are-seeking-way-grow-their-money-barbarians-farm-gate

The Economist (2015b) The toll of a tragedy. http://www.economist.com/blogs/ graphicdetail/2015/03/ebola-graphics

The IUCN Redlist of Threatened Species (2016) http://www.iucnredlist.org/

The Telegraph (2015) Airbus's quantum computing brings Silicon Valley to the Welsh Valleys. 26 December 2015, http://www.telegraph.co.uk/finance/newsby-sector/industry/12065245/Airbuss-quantum-computing-brings-Silicon-Valley-to-the-Welsh-Valleys.html

The World Bank (2013) Turn down the heat. http://www.worldbank.org/en/news/ feature/2013/06/19/india-climate-change-impacts

The World Counts (2015) Hazardous waste statistics. The World Counts, Copenhagen. http://www.theworldcounts.com/counters/waste_pollution_facts/ hazardous_waste_statistics

Theodore T (2010) Optical illusion a possible factor in deadly plane crash. News Vancouver, 1 October 2010, http://bc.ctvnews.ca/optical-illusion-a-possible-factors-in-deadly-plane-crash-1.558708

Thomas CD et al (2004) Extinction risk from climate change. Nature 427:145–148. doi:10.1038/nature02121. http://www.nature.com/nature/journal/v427/n6970/abs/nature02121.html

Thornton JW et al (2002) Biomonitoring of industrial pollutants. Public Health Reports, July

Times of India (2010) Groundwater in 33% of India undrinkable. Times of India, 12 March 2010, http://timesofindia.indiatimes.com/india/Groundwater-in-33-of-India-undrinkable/articleshow/5673304.cms?referral=PM

Torres P (2016) Biodiversity loss: an existential risk comparable to climate change. The Bulletin of the Atomic Scientists. http://thebulletin.org/biodiversity-loss-existential-risk-comparable-climate-change9329#.VxRxznC6AAw.twitter. Accessed 11 Apr 2016

Tribe D (1994) Feeding and greening the world. CAB International, Oxford

Turco R, Toon B, Ackerman T, Pollack J, Sagan C (1983) Nuclear winter: global consequences of multiple nuclear explosions. Science 222:1283–1292

Turco R et al (2012) The climatic effects of nuclear war. Chapter 4.1 in Lights out: how it all ends. Sci Am 2012

Tuttle B (2012) Got stuff? Typical American Home is cluttered with possessions—and stressing us out. Time Magazine, July 19, 2012

UDHR (2016) The Universal Declaration of Human Rights. United Nations website, http://www.un.org/en/documents/udhr/

UN ESA (2014) http://esa.un.org/unpd/wpp/

UN FAO (2013) How to feed the world in 2050, 2013. http://www.fao.org/fileadmin/templates/wsfs/docs/expert_paper/How_to_Feed_the_World_in_2050.pdf

UN FAO (2014a) The case for energy-smart food systems. http://www.fao.org/docrep/014/i2456e/i2456e00.pdf

UN FAO (2014b) The state of world fisheries and aquaculture (SOFIA) 2014, www.fao.org/3/a-i3720e.pdf

UN Department of Economic and Social Affairs (DESA) (2014) World's population increasingly urban with more than half living in urban areas. http://www.un.org/en/development/desa/news/population/world-urbanization-prospects-2014.html

US Energy Information Administration (2013)

US Energy Information Administration (2015a) International energy statistics. A Btu is the amount of energy it takes to raise a pint of water one degree Fahrenheit in temperature. http://www.eia.gov/cfapps/ipdbproject/IEDIndex3.cfm?tid=44&pid=44&aid=2

US Energy Information Administration (2015b) http://www.eia.gov/countries/index.cfm?view=production

UNCCD (2014) Syria's conflict: is land degradation part of the story? UNCCD News 6.1. p 5. http://newsbox.unccd.int/imgissue/UNCCDNews6_1.pdf

UNEP (2002) A threat to natural resources. http://www.grid.unep.ch/waste/html_file/16-17_consumption_threat.html

UNEP (2007) Forest losses and gains: where do we stand? http://www.unep.org/vitalforest/Report/VFG-02-Forest-losses-and-gains.pdf

UNEP (2015) Extrapolated from world minerals production statistics. http://www.grid.unep.ch/waste/html_file/16-17_consumption_threat.html

UNEP Centre for Clouds, Chemistry and Climate (2002) The Asian brown cloud. Also The Economist, 10 August 2013 'The east is grey'. www.economist.com/news/briefing/21583245-china-worlds-worst-polluter-largest-investor-green-energy-its-rise-will-have

UNFAO (2015a) How soil is destroyed. http://www.fao.org/docrep/T0389E/T0389E02.htm

UNFAO (2015b) Global forest resources assessment 2015. How are the world's forests changing? http://www.fao.org/3/a-i4793e.pdf. Accessed 7 Sept 2015

UNFAO (2015c) The post-2015 development agenda and the millennium development goals: fisheries, aquaculture, oceans and seas. http://www.fao.org/post-2015-mdg/14-themes/fisheries-aquaculture-oceans-seas/en/

United Nations (2015) Desertification. http://www.un.org/en/events/desertificationday/background.shtml

United Nations Environment Program (2013) Global chemicals outlook: towards sound management of chemicals, February 2013, http://www.unep.org/chemicalsandwaste/Portals/9/Mainstreaming/GCO/The%20Global%20Chemical%20Outlook_Full%20report_15Feb2013.pdf

United Nations Population Division (2012) World population prospects: the 2012 revision. https://data.un.org/Data.aspx?d=PopDiv&f=variableID%3A54

University of NSW (2012) Breakthrough in bid to create first quantum computer, UNSW Press Release, 20 September 2012. http://newsroom.unsw.edu.au/news/technology/breakthrough-bid-create-first-quantum-computer

University of Toronto (2013) Guidelines on the use of perfumes and scented products. University of Toronto Environmental Health and Safety. www.ehs.utoronto.ca/resources/HSGuide/Scent.htm#Purpose

UN-REDD (2008) About the UN-REDD Programme 2008–2016. http://www.un-redd.org/AboutUN-REDDProgramme/tabid/102613/Default.aspx

UN Food and Agriculture Organisation (2014) http://www.un.org/waterforlifedecade/food_security.shtml

US EPA (2013) EPA Releases Report Containing Latest Estimates of Pesticide Use in the United States, February 17, 2011. http://epa.gov/oppfead1/cb/csb_page/updates/2011/sales-usage06-07.html

US CDC (2014) Fourth National Report on Human Exposure to Environmental Chemicals, August 2014. http://www.cdc.gov/exposurereport/pdf/fourthreport_updatedtables_aug2014.pdf

US Department of Health and Human Services (2014) National Toxicology Program, December 2014. http://ntp.niehs.nih.gov/about/index.html

US Federal Bureau of Investigation (2016) Famous criminals or cases: amerithrax or anthrax investigation. FBI website, http://www.fbi.gov/about-us/history/famous-cases/anthrax-amerithrax/amerithrax-investigation. Accessed 2016

US National Academy of Sciences (2015) Climate intervention is not a replacement for reducing carbon emissions; proposed intervention techniques not ready for wide-scale deployment. http://www8.nationalacademies.org/onpinews/newsitem.aspx?RecordID=02102015

USEPA (2014) Protecting the stratospheric ozone layer, October 2014, http://www.epa.gov/airquality/peg_caa/stratozone.html

Van der Gun J (2012) Groundwater and global change: trends, opportunities and challenges. UNESCO

Van Trump K (2015) Percent of income Americans spend on food half what it was in 1960. The Van Trump Report (blog), 5 March 2015. http://vantrumpreport.com/786/?utm_content=buffer74b43&utm_medium=social&utm_source=twitter.com&utm_campaign=buffer

van Wyck B (2013) The groundwater of 90% of Chinese cities is polluted. Danwei, 18 February 2013, www.danwei.com/the-groundwater-of-90-of-chinese-cities-is-polluted

Vidal J (2010) UN report: world's biggest cities merging into 'mega-regions'. The Guardian. http://www.theguardian.com/world/2010/mar/22/un-cities-mega-regions

Voiland A (2014) Earth's disappearing groundwater. NASA. November 2014. http://earthobservatory.nasa.gov/blogs/earthmatters/2014/11/05/earths-disapearing-groundwater/?utm_content=bufferf656a&utm_medium=social&utm_source=twitter.com&utm_campaign=buffer

Wada Y (2012) Non-sustainable groundwater sustaining irrigation. Global Water Forum. February 2012. http://www.globalwaterforum.org/2012/02/13/non-sustainable-groundwater-sustaining-irrigation/

Wang L (2015a) World's largest indoor vertical farm will produce 2 million pounds of soil-free food in Newark. Inhabitat. 12 March 2015. http://inhabitat.com/worlds-largest-indoor-vertical-farm-will-produce-2-million-pounds-of-soil-free-food-in-newark/

Wang L (2015b) World's largest vertical indoor farm. Inhabitat. 11 March 2015. http://inhabitat.com/worlds-largest-indoor-vertical-farm-will-produce-2-million-pounds-of-soil-free-food-in-newark/

Ward PD (2007) Under a Green Sky. Smithsonian Books. For a recent summary of Permian Extinction theory see Monbiot G, How fossil fuel burning nearly wiped out life on Earth—250m years ago. The Guardian, May 27 2015. http://www.theguardian.com/commentisfree/2015/may/27/threat-islamic-state-fossil-fuel-burning?CMP=share_btn_link

Ward P (2008) Under a green sky. Harper Perennial, New York

Waskow A (2015) Torah, pope, & crisis inspire 300+ rabbis to call for vigorous climate action. The Shalom Centre, June 8, 2015. https://theshalomcenter.org/torah-pope-crisis-inspire-300-rabbis-call-vigorous-climate-action

Water Footprint Network (2015) http://www.waterfootprint.org/?page=files/ YourWaterFootprint

Weber M (2008) The 'good myth' of world war II. Institute for historical review. http://www.ihr.org/news/weber_ww2_may08.html

Weldon C, du Preez LH, Hyatt AD, Muller R, Speare R (2004) Origin of the amphibian chytrid fungus. Emerg Infect Dis. http://wwwnc.cdc.gov/eid/ article/10/12/03-0804, December 2004

Whitehouse S (2014) The Climate Denial Beast, US Senate. 4 February 2014, http:// www.whitehouse.senate.gov/news/speeches/the-climate-denial-beast

Whitley S (2013) Time to change the game: fossil fuel subsidies and climate. Overseas Development Institute, November 2013. http://www.odi.org/sites/odi.org.uk/ files/odi-assets/publications-opinion-files/8669.pdf

WHO (2012) State of the science of endocrine disrupting chemicals. http://www. who.int/ceh/publications/endocrine/en/

WHO (2014) Urban population growth. http://www.who.int/gho/urban_health/ situation_trends/urban_population_growth_text/en/

WHO (2016) Preventing disease through healthy environments: a global assessment of the burden of disease from environmental risks. World Health Organisation, 15 March 2016. http://www.who.int/mediacentre/news/releases/2016/deaths-attributable-to-unhealthy-environments/en/#.Vuj3krzjMgY.email

Wikipedia (2015) Syria and weapons of mass destruction. http://en.wikipedia.org/ wiki/Syria_and_weapons_of_mass_destruction

Wilcox C, Van Sebille E, Hardesty BD (2015) Threat of plastic pollution to sea-birds is global, pervasive. Proc Natl Acad Sci. http://www.pnas.org/content/ early/2015/08/27/1502108112, August 28

Wildshutte JH et al (2016) Discovery of unfixed endogenous retrovirus insertions in diverse human populations. PNAS, 11 February 2016, http://www.pnas.org/ content/early/2016/03/16/1602336113

Wilkinson BH, McElroy BJ (2006) The impact of humans on continental erosion and sedimentation. Geological Society of America Bulletin, July 2006, http:// gsabulletin.gsapubs.org/content/119/1-2/140.abstract

Wilkinson BH, McElroy BJ (2007) The impact of humans on continental erosion and sedimentation. Geol Soc Am Bull 119(1/2):140–156. doi:10.1130/B25899.1

Wilson EO (2016a) The global solution to extinction. New York Times, 12 March 2016. http://www.nytimes.com/2016/03/13/opinion/sunday/the-global-solution-to-extinction.html?_r=0. Also https://www.youtube.com/watch?v=7ANire8E240

Wilson EO (2016b) Half Earth: our planet's fight for life. W.W. Norton & Company

WIN-Gallup International (2012) Global index of religiosity and atheism. http:// www.wingia.com/web/files/richeditor/filemanager/Global_INDEX_of_ Religiosity_and_Atheism_PR__6.pdf

Winkelmann R, Levermann A, Ridgwell A, Caldeira K (2015) Combustion of available fossil fuel resources sufficient to eliminate the Antarctic Ice. Sci Adv 1(8):e1500589. doi:10.1126/sciadv.1500589 http://advances.sciencemag.org/

content/1/8/e1500589.full and http://e360.yale.edu/feature/rising_waters_how_fast_and_how_far_will_sea_levels_rise/2702/

WMO (2016) 2015 is the hottest year on record. 25 January 2016, https://www.wmo.int/media/content/2015-hottest-year-record

Wolchover N (2011) How many people can earth support?, LiveScience. 11 October 2011. http://www.livescience.com/16493-people-planet-earth-support.html

Wood C (2006) The Dutch tulip bubble of 1637. Damn Interesting. http://www.damninteresting.com/the-dutch-tulip-bubble-of-1637/

Woolf N (2014) Ebola isn't the big one. So what is? And are we ready for it? The Guardian, 3 October 2014. http://www.theguardian.com/world/2014/oct/03/-sp-ebola-outbreak-risk-global-pandemic-next

World Bank (2012) New report examines risks of 4 degree hotter world by end of century. World Bank, November 18, 2012. http://www.worldbank.org/en/news/press-release/2012/11/18/new-report-examines-risks-of-degree-hotter-world-by-end-of-century

World Bank (2013) Turn down the heat: climate extremes, regional impacts and the case for resilience. World Bank, June 2013. http://www.worldbank.org/content/dam/Worldbank/document/Full_Report_Vol_2_Turn_Down_The_Heat_%20Climate_Extremes_Regional_Impacts_Case_for_Resilience_Print%20version_FINAL.pdf

World Bank (2015a) CO_2 emissions. http://data.worldbank.org/indicator/EN.ATM.CO2E.PC

World Bank (2015b) GINI index (World Bank estimate), http://data.worldbank.org/indicator/SI.POV.GINI/

World Bank (undated) Waste generation. Urban Development Series, http://siteresources.worldbank.org/INTURBANDEVELOPMENT/Resources/336387-1334852610766/Chap3.pdf

World Coal Association (2013) Coal statistics. http://www.worldcoal.org/resources/coal-statistics/

World Economic Forum (2015) Global risks 2015

World Fair Trade Organization (2016) 10 principles of fair trade. http://www.wfto.com/fair-trade/10-principles-fair-trade

World Footprint Network (2016) http://www.footprintnetwork.org/en/index.php/GFN/

World Glacier Monitoring Service (2015) Global glacier changes: facts and figures. http://www.grid.unep.ch/glaciers/pdfs/5.pdf

World Health Organization (2012) List of countries by life expectancy. http://en.wikipedia.org/wiki/List_of_countries_by_life_expectancy

World Health Organization (2014) The top 10 causes of death. WHO, May 2014. http://www.who.int/mediacentre/factsheets/fs310/en/

World Health Organization (2015a) Pandemic and epidemic diseases. http://www.who.int/csr/disease/en/

World Health Organization (2015b) Anticipating epidemics. http://www.who.int/csr/disease/anticipating_epidemics/en/

World Health Organization (2015c) Global alert and response: smallpox. http://www.who.int/csr/disease/smallpox/en/

World Meteorological Organization (1988) Conference Proceedings. The Changing Atmosphere: Implications for Global Security, Toronto, Canada, 27–30 June 1988. Secretariat of the World Meteorological Organization, Geneva 1989

World Preservation Foundation (2016) http://www.worldpreservationfoundation.org/topic.php?cat=climateChange&vid=48#.VGgDLVJxmqE and Howard BC. 8 Mighty Rivers Run Dry from Overuse, National Geographic, http://environment.nationalgeographic.com.au/environment/photos/rivers-run-dry/

World Resources Institute (2015) Natural Resources: what are they?

World Trade Center Health Program (2015) 6th Meeting of the Scientific/Technical Advisory Committee, 4 June 2015. http://www.cdc.gov/wtc/pdfs/WTCHealthProgram-STACMeetingJune42015-PAReviewed.pdf

Worldwide Fund for Nature (WWF) (2015) Living Blue Planet Report 2015. http://assets.worldwildlife.org/publications/817/files/original/Living_Blue_Planet_Report_2015_Final_LR.pdf?1442242821&_ga=1.160971965.149081568.1442451626

Wu M et al (2010) Case report: lung disease in World Trade Center responders exposed to dust and smoke: carbon nanotubes found in the lungs of World Trade Center patients and dust samples. Environ Health Perspect 118(4):499–504. http://www.ncbi.nlm.nih.gov/pmc/articles/PMC2854726/

WWF (2014) Living Planet Report 2014. http://wwf.panda.org/about_our_earth/all_publications/living_planet_report/

Yeo B, Langley-Turnbaugh S (2010) Trace element deposition on Mt Everest. Soil Survey Horizons 51(4):95–101

Youssef M (2010) Penticton plane crash caused by excess weight, probe finds. The Globe and Mail, 1 October 2010. http://www.theglobeandmail.com/news/british-columbia/penticton-plane-crash-caused-by-excess-weight-probe-finds/article4389785/

Zeebe RE, Zachos JC, Dickens GR (2009) Carbon dioxide forcing alone insufficient to explain Palaeocene–Eocene Thermal Maximum warming. Nat Geosci 2:576–580. doi:10.1038/ngeo578. http://www.nature.com/ngeo/journal/v2/n8/full/ngeo578.html#B5

Zemp M et al (2015) Historically unprecedented global glacier decline in the early 21st century. J Glaciol 61(228). doi: 10.3189/2015JoG15J017

Index

© Springer International Publishing Switzerland 2017
J. Cribb, *Surviving the 21st Century*, DOI 10.1007/978-3-319-41270-2

The Author

Julian Cribb is an Australian author and science writer. A former newspaper editor and science communicator his published work includes over 8000 articles, 3000 science media releases and nine books. He has received more than 30 awards for journalism and is a fellow of the Australian Academy of Technology, Science and Engineering and the Australian National University Emeritus Faculty.

His internationally-acclaimed book, *The Coming Famine* (University of California Press, 2010) explored the question of how we can feed 10 billion humans this century. His book *Poisoned Planet* (Allen and Unwin 2014) investigates the contamination of the Earth system and all humanity by man-made chemicals. *Surviving the 21st Century* is the third book in his trilogy about how humans can overcome the existential risks our success has brought upon us.

© Springer International Publishing Switzerland 2017
J. Cribb, *Surviving the 21st Century*, DOI 10.1007/978-3-319-41270-2